FIFTH EDITION

MUSCLE TESTING
TECHNIQUES OF MANUAL EXAMINATION

LUCILLE DANIELS, M.A.
Professor of Physical Therapy, Emerita,
School of Medicine, Stanford University

CATHERINE WORTHINGHAM, PH.D., D.SC.
Formerly Director of Professional Education,
The National Foundation, Inc.

W. B. SAUNDERS COMPANY
Philadelphia, London, Toronto, Mexico City, Rio de Janeiro, Sydney, Tokyo, Hong Kong

DANIELS   1854-5

W.B. SAUNDERS COMPANY
A Division of
Harcourt Brace & Company

The Curtis Center
Independence Square West
Philadelphia, Pennsylvania 19106

Library of Congress Cataloging-in-Publication Data

Daniels, Lucille.
  Muscle testing.
  Bibliography: p.
  Includes index.

  1. Muscles—Examination. I. Worthingham, Catherine.
II. Title. [DNLM: 1. Muscles—physiology. 2. Physical
                Examination—methods. WE 500 D186m]
RC925.7.D36   1986   616.7'40754   85-22220
ISBN 0-7216-1854-5

Editor: Baxter Venable
Production Manager: Frank Polizzano
Manuscript Editor: Edna Dick

Listed here is the latest translated edition of this book with the language of the translation and the publisher.

French **(4th Edition)**—Maloine S.A. Editeur, Paris, France
German **(4th Edition)**—Gustav Fischer Verlag, Stuttgart-Hohenheim, Germany
Italian **(3rd Edition)**—Verduci Editore, Rome, Italy
Japanese **(4th Edition)**—Kyodo Isho Shuppan Sha, Tokyo, Japan
Spanish **(4th Edition)**—Nueva Editorial Interamericana, S.A., de C.V., Mexico
Greek **(3rd Edition)**—Gregory Parisianos, Athens, Greece
Portuguese **(4th Edition)**—Editora Interamericana Ltda., Rio de Janeiro, Brazil

Muscle Testing   ISBN   0-7216-1854-5

Last digit is the print number: 18   17   16   15   14   13   12   11   10

PREFACE TO THE FIFTH EDITION

Manual muscle tests are used to determine the extent and degree of muscular weakness resulting from disease, injury or disuse. The records obtained from the tests provide a basis for planning therapeutic procedures and for periodic retesting, which can be utilized in evaluating these procedures. Muscle testing is therefore an important tool for all members of the health team dealing with the physical residual of disability.

The public and the health professionals are demanding a total approach to patient treatment. The emphasis has changed from "sick care" to "health care" through the pursuit of comprehensive, continuous and coordinated care. Consequently physicians, physical therapists, occupational therapists, nurses and their supporting personnel have need for varying levels of knowledge and ability in muscle testing in their concern for preventing disability, habilitating those who have never known normal function, restoring those with disability to optimal function and maintaining the function obtained. Physical educators, although less concerned with the treatment of muscle impairment, are definitely concerned with the optimal development of the body and the prevention of disability. They must, therefore, be familiar with the details of muscle function. Testing for fair, good and normal function is an excellent device for teaching kinesiology to this group.

In this, the fifth edition, a number of changes have been made. Measurement of the ranges of joint motion has been added; the illustrations have been revised and the terminology up-dated. The explanatory material for the tests has been expanded, particularly in relation to muscle substitution.

We wish to thank the members of the faculties of the educational programs for physical therapists who have given us suggestions for the revisions. Our special appreciation goes to Ann Hallum and Barbara Kent from Stanford for their support.

LUCILLE DANIELS
CATHERINE WORTHINGHAM

CONTENTS

INTRODUCTION

The technique of manual muscle examination presented in this text is based on the work of a number of investigators. No attempt has been made to present all the tests devised for any particular muscle or muscle group. As the brain thinks in terms of movement rather than the contraction of an individual muscle, the emphasis has been placed on the testing of prime movers in relation to the principal joints of each segment of the body. The prime mover (or movers) may be one or more than one muscle.

The analysis of gait is presented as a screening device for the muscle testing of the ambulatory patient. The relationship between deviations in the patient's walking pattern and possible muscle weakness or limitation in the range of motion is outlined. The results of the gait analysis may be used to validate the test grades, and the grades may verify the results of the analysis when weakness or joint limitation is present.

Procedures used in measuring the ranges of joint motion have been added, as limited motion is commonly associated with muscle weakness. Inability to complete a range of motion should be noted on the patient's record and the degrees of limitation recorded.

DEVELOPMENT OF MANUAL MUSCLE TESTING

Dr. Robert W. Lovett, Professor of Orthopedic Surgery at Harvard Medical School, was the originator of the gravity tests. Janet Merrill, Director of Physical Therapeutics at Children's Hospital and the Harvard Infantile Paralysis Commission in Boston, and an early coworker of Dr. Lovett, stated that the tests were first used in his office gymnasium in 1912.* The first published article describing the procedures, which included the use of outside force, was written by Wilhelmina Wright, who assisted Dr. Lovett.

In 1922, Charles L. Lowman, an orthopedic surgeon in Los Angeles, worked out a numerical system for grading muscle action. It was used with the gravity tests but covered the ranges of active motion in the joints in much greater detail, particularly when extreme muscular weakness was present.

A percentage system of recording was originated in 1936 by Henry O. and Florence P. Kendall, physical therapists at the Children's Hospital School in Baltimore. This system was based on a recording range in the gravity and resistance tests from 0 to 100 per cent and introduced the element of fatigue in grading. The work of these authors, which was first published in a United States Public Health Service bulletin, has been widely used.

In 1940 Signe Brunnstrom and Marjorie Dennen, physical therapists from the Institute for the Crippled and Disabled in New York City, prepared a syllabus on muscle testing. A detailed system of grading movements rather than individual muscles was outlined. This was an adaptation of the Lovett gravity and resistance tests, and included the use of fatigue as an element in grading.

In 1940, Elizabeth Kenny from Australia introduced a system for recording the presence of function, spasm and incoordination in muscles affected by poliomyelitis, which was called a "muscle analysis." This was one of the many significant contributions to the treatment of patients with poliomyelitis that she made in her homeland and in this country. In 1942, Alice Lou Plastridge, Director of Physical Therapy, Georgia Warm Springs Foundation, correlated the use of the "muscle analysis" with manual muscle testing. The analysis was used primarily during the acute stage of poliomyelitis and served to supplement the strength tests in the convalescent and chronic stages.

Beginning with 1951, manual muscle testing played a vital part in the evaluation of agents designed to combat paralytic poliomyelitis. The first field trial studies were carried out to determine whether gamma globulin would protect against paralysis resulting from poliomyelitis with the support of the National Foundation for Infantile Paralysis. Muscle tests were given by physical therapists in three

*Letter to Lucille Daniels from Janet Merrill, January 5, 1945.

epidemic areas selected for the trials. Grading was done by the gravity and manual resistance techniques, using the Lovett system of grading. To assess involvement, a numerical method was devised by Dr. Jessie Wright and her associates at the D. T. Watson School of Physiatrics, Leetsdale, Pennsylvania. The muscle grades were given numerical values, and each muscle or group was assigned an arbitrary factor according to its bulk. The factor multiplied by the muscle grade resulted in an "index of involvement" expressed in a percentage figure.

In the 1954 field trial of poliomyelitis vaccine, 67 physical therapists participated, using the abridged form of the record previously developed.

In 1961, Smith, Iddings, Spencer, and Harrington reported the development of a numerical index for clinical research in muscle testing. A detailed form was used that included the addition of + and – to the standard grades. The authors noted the importance of a numerical index of total involvement, not as a substitute for the customary muscle test but to make it more useful in the areas of education and research. A further report of Iddings, Smith, and Spencer, published the same year, discussed reliability in the clinical use of muscle testing. The results indicated that these tests can be highly reliable in spite of variation in the educational preparation of physical therapists and the use of different techniques of manual muscle testing. The average difference in grading among all physical therapists in the study was approximately 4 per cent, which compared favorably with the 3 per cent found in the more abridged form of testing in the poliomyelitis field trials.

In recent years numerous mechanical and electronic devices of varying complexity and applicability to clinical muscle testing have been developed. These provide valuable information concerning muscle function and are important for further research. However, manual muscle testing remains a readily available and inexpensive tool for both clinical and research purposes.

BASIC CONSIDERATIONS IN TESTING

Validity and Reliability

Careful observation, palpation, stabilization and correct positioning are essential for validity in testing. The patient should be asked to attempt to move the part through the range of motion, if he is able to do so. The examiner should observe and note dissimilarities in the size and contour of the muscle or group of muscles being tested and the counterpart on the opposite side of the body. The contractile tissue and tendon (or tendons) should be palpated, as a lack of tension helps to identify substitution by muscles other than the prime movers. Substitution usually can be eliminated by stabilization and careful positioning; if this is impossible, a notation should be made on the record. A classic example of complete substitution can occur in patients with muscular dystrophy, when the prime movers may be nonfunctioning and secondary muscles perform the movement.

In considering the interpretation of a test grade it is probably unnecessary to point out the existence of variation in length and bulk of the body parts, variations in shape of like parts in different persons, age and sex differences in strength, and the ever-present psychological considerations of cooperation and willingness to put forth maximal effort that operate particularly in very young children. In view of these and other factors such as fatigue, it would be an error to assume that invariably muscles or muscle groups with the same grade have suffered an equal degree of involvement.

A factor that has been widely disregarded, not only in muscle testing but also in therapeutic exercise in general, is the considerable variation in the force that a muscle can normally exert at various points through the range of motion of the moving segment. Consideration of such "strength curves" or "isometric joint torque curves" shows that the test point in manual resistance tests is often near the weakest portion of the range. As long as the test is always done in the same manner, this will not affect its reliability, but it may have implications for functional interpretation of the grade.

The Grading System

The basic grades used throughout the following pages are based on three factors:

1. The amount of resistance that can be given manually to a contracted muscle or muscle group (normal or good grade);

2. The ability of the muscle or muscle group to move a part through a complete range of motion against gravity (a vertical movement) for a Fair grade or with gravity decreased (a horizontal movement) for a Poor grade. When horizontal movements are impractical, a partial range on the vertical plane is substituted; and

3. Evidence of the presence or absence of a contraction in a muscle or muscle group (slight contraction without joint motion—a trace grade, no contraction—a zero).

In addition to the use of the basic grades, a common practice is to add a plus or minus to denote:

1. A greater or lesser amount of resistance than the normal or good grades signify (slightly less resistance than the amount that can be given to a normal muscle —N−, or slight resistance at the end of the range of motion against gravity −F+);

2. A variation in the range of motion that may be graded fair or poor (range of motion can be completed with gravity decreased and also a partial range against gravity −P+). The use of *plus* or *minus* in the resistance tests is based on a subjective decision made by the examiner. In the gravity tests, a division in the range of motion may be used that increases the objectivity of the evaluation. If less than half of the range is completed, the lesser grade with a plus is recorded; if more than half but not the full range, the higher grade with a minus is used (e.g., P+ and F− respectively for motion against gravity).

The complication of limited passive motion is of importance. The range-grade system that requires the recording of the degrees of motion with the grade is often used. For example, if the passive range in elbow flexion is limited to 90 degrees, and the patient can complete this range against gravity, it is recorded as 0°—90°/F.

Normal and Good Grades

The amount of resistance required for a grade of normal or good varies with the individual patient and the muscle or muscle group examined. If the muscles in the contralateral side of the body are known to be uninvolved, valid information can be obtained by giving resistance to each counterpart before testing the involved muscles. Otherwise, the examiner must depend upon previous experience to make a judgment.

Resistance given at the end of the range of motion (break test) to determine good and normal grades is simpler and can be applied more quickly than resistance throughout the range of action. However, many clinicians prefer to give resistance through the range before the break test in order to sense the amount of power being generated by the muscle or muscle group by the contraction. In the break test, the patient can usually follow the request to "hold" with ease; however, be sure there is time to establish a maximum contraction before resistance is given. Pressure should always be exerted in a direction as nearly as possible opposite to the line of pull of the muscle or muscle group being tested, and at the distal end of the segment on which it inserts.

Pain should not be elicited in a break test. Gradual, increasing pressure should be given and the patient observed closely for any evidence of discomfort or pain, and resistance should be discontinued if either occurs.

Fair Grade

The ability to raise a segment through its range of motion against gravity appears to be a fairly specific accomplishment, lying somewhere between the extremes of not being able to contract the muscle and holding the segment at the end of its range of motion against "normal" maximal resistance. Manual muscle testing in its simplest form centers on this concept, with reliance on the judgment and skill of the examiner to determine whether the muscle or group being tested is at a "fair" point in performance, or is above or below it, and to what degree.

It may appear that a direct comparison of fair grades is reasonable, since larger parts have greater muscle forces available to move them. This is true to some extent, although surprising variations exist in the ratio of the weight of the part and the maximum force normally available to lift it. For example, direct force measurements have shown that with the subject supine and the head relaxed in a sling, gravity exerts a downward force on the head that has been recorded as 9 pounds. Upward force resulting from maximum contraction of normal neck flexors may be 19 pounds, as measured by means of the sling, or a total of 28 pounds, including the force required to support the head. Thus, in this instance, the ratio of fair to normal is about 9:28, or 32 per cent. In contrast, with the subject seated the resistance of the relaxed forearm supported in a horizontal position by a test strap at the wrist may be 5 pounds and maximum elbow flexor contraction may result in an upward force of 75 pounds measured at the wrist. The ratio of these two values is 5:80, or 6.3 per cent. (Similar measurements have shown the ratio for the quadriceps to be 8:80, or 10 per cent, in some instances, and for the hip abductors 12:50, or 24 per cent). Actually, such measurements involve force moments rather than true muscle forces and segmental weights, but if the lever lengths of downward and upward forces are kept equal, they may be ignored in estimating the ratios of the two.

The few figures cited should not be interpreted as typical of large numbers of persons, since there are undoubtedly wide variations according to age, body configuration, and other factors as well as variations in the manner of giving the dynamometer tests. These figures are offered here to show the hazards of assigning arbitrary numerical values to the original Lovett grades, which may lead to misinterpretation. Numerals are acceptable for the recording of muscle performance only if the concepts of the grades and the tests are kept in perspective.

Direct force measurements show that the level of fair is usually relatively low, so that a vastly greater range exists between this point and normal than between this point and a trace.

A fair grade might be said to represent *a definite functional threshold* for each individual movement tested, indicating that the muscle or muscles can achieve the minimum task of moving the part upward against gravity through its range of motion. Though this ability is significant for the upper extremity, it falls far short of the functional requirements for many of the lower extremity muscles used in walking, particularly for such groups as the hip abductors, knee extensors, and plantar flexors and dorsiflexors of the foot.

Poor Grade

The poor grade denotes the patient's ability to move a part through a range of motion on the horizontal plane, which can be termed a gravity-decreased movement. Exceptions are tests for the fingers and toes in which the weight of the parts is not significant and tests for which the gravity-decreased positions are not practical, e.g., flexion and extension of the neck. (This test can be given with the patient sidelying, head supported by the examiner's hand. However, the examiner will find it quite difficult to refrain from assisting the patient in completing the range of motion. For this reason, the test is not considered to be practical.) For these, a partial range against gravity may be graded poor and a full range fair.

Although considered below the functional range, muscles graded poor provide a measure of stability to a joint that is of value to the patient. It should also be noted that the identification of this level of function is important in the early stages of a disability as a muscle with a poor grade has a greater potential for an increase in strength than one that receives a lower grade of trace or zero.

Trace and Zero Grades

A trace or an absence of a muscle contraction is determined by careful observation and palpation of the tendons and the muscle bulk. An increase in tension or a flicker of movement may be more easily seen or palpated in a tendon if near the surface of the body. These should be checked first, followed by inspection and palpation of the contractile tissue. It is difficult, and sometimes impossible, to identify a minimal contraction in one of the deep muscles of the body. Usually it cannot be done unless the overlying muscles are nonfunctioning, and the contraction of the muscle being tested is sufficient for the line of pull to be identified. Recording a trace or zero with a question mark may be indicated.

Stabilization

Manual stabilization is used in testing for adequate fixation to isolate the desired action to a specific joint. A muscle in contracting pulls on its origin as well as its insertion and with equal force. To obtain maximum muscle action, the stationary segment, which in testing is usually the site of the origin, must be fixed by muscle tension, gravitational pull, or external pressure from manual stabilization. Therefore, care must be taken that muscles are not placed at a disadvantage by failure of the fixator force and consequently penalized in grading.

Synergistic action refers to a contraction of all the muscles acting around a joint. These include the prime movers, the muscles that act in concert with the prime movers to define the spatial limits, and the antagonists that check or limit the movement. For example, the long finger flexors in flexing the phalanges of the fingers with maximum tension would also flex the wrist if the wrist extensors did not prevent this. In testing, the necessity for this type of muscle synergy is ordinarily eliminated by the stabilization applied by the examiner during the test.

Limitations of Manual Muscle Testing

The muscle testing methods presented in this text were designed for use in assessing the extent and degree of weakness following disorders primarily involving the contractile muscle elements, the myoneural junction, and the lower motor neuron. Disorders that affect the organization of movement in the higher levels of the central nervous system, such as cerebral palsy or hemiparesis secondary to cerebral vascular accident, alter reflex activity and bring about changed states of tone in complete muscle synergies. Although muscle weakness exists, an evaluation by voluntary movements in the selected positions outlined in this book will be misleading. Methods now are available for assessment of relative degrees of weakness (hypotonia) and hypertonia in synergistic muscle groups by altering limb position and total body posture. A review of these methods, however, is not within the scope of this book.

Screening Tests

The examiner's time must be conserved, and the patient's fatigue must be considered in a detailed muscle examination; therefore, screening tests have been found to have a useful function. In one procedure, the part is placed passively (by the examiner) in the position used for the normal test without regard for gravity. If the patient can hold against resistance, a judgment is made by the examiner in regard to a normal or good grading. If the patient cannot hold against resistance, use of the standard tests for determining grades below good is indicated.

Another screening procedure is to combine tests for the extremities, such as checking the flexors or abductors of both shoulders simultaneously in the sitting position, and the abductors or adductors of both hips in the supine position.

With experience, the examiner will be able to devise many "quickie" tests, particularly for patients who appear to have generalized weakness. An example is the hand grasp (handshake position) in which the strength of the finger and thumb flexors can be determined by the amount and evenness of the pressure exerted by each phalanx against the examiner's hand. The muscles controlling wrist motion may also be tested by resistance given in this position while the examiner stabilizes the forearm. In an additional test for the hand the examiner approximates the palmar surface of his hand to the dorsal surface of the patient's and gives resistance to all the finger and thumb extensors simultaneously.

Careful observation of patients performing ordinary activities will often provide clues to poor function and is an important part of the evaluation procedure. With health care being offered to an increasing number of geriatric patients, screening tests have become important tools for those members of the health services concerned with the welfare of these patients. With experience, the accuracy of the screening tests will be increased as the examiner develops the ability to discern not only the gross but the less obvious deviations from the normal patterns of movement. Valid judgments of the functional level may be formed without expending excessive time or unduly tiring the patients.

Gait Analysis

The analysis of gait is based on meticulous observation of the ambulatory patient during standing and walking. Deviations from the normal stance of the patient that may affect gait are noted first, followed by the abnormalities in both the general and specific elements of the walking cycle. These deviations identify the areas of weakness or other factors limiting normal function. Using the clues as a guide, the examiner may then proceed to carry out such tests as are indicated. The recorded data from a gait analysis may also be used for periodically determining the degree of improvement in the basic functional activity. Movies or film strips of a patient's walking cycle are also valuable records that can be used for this purpose.

A detailed consideration of the screening procedures used in gait analysis is presented in the last section of this publication following the standard muscle tests.

Goniometric Measurements

The goniometer is an instrument used to measure the angle in a range of joint motion limited by disease, injury or disuse. The record is of value in determining the extent of disability, the future care, and, when repeated at intervals, the effectiveness of procedures used to obtain normal motion in the rehabilitation process. Such objective evidence is of value to all persons involved in the patient's treatment and to the patient as a motivating factor. External groups such as insurance companies, health agencies, and those in the judicial system can more readily interpret the condition of a patient with joint limitations when an accurate record of measurements is available.

The technique of measurement that has been added to this edition involves the joints of the hips, shoulders and extremities. No attempt has been made to include the spine, as the goniometer is not well suited to the measurement of multiple joints. A tape measure often is used to record limitation in movements of the trunk, neck and head.

The system of measurement used in this publication is based on the half circle, 0 to 180 degrees. The anatomical position of the body has been established as the zero point with the exception of hip and shoulder rotation, and forearm supination and pronation. After a joint with limited motion is measured, if the one on the contralateral side of the body is uninvolved, it should be measured to determine the normal range for the individual patient.

Note that no axis (fulcrum) is given in the measurement of each joint, except when special placement of the goniometer is required. Otherwise, if the two arms are aligned properly on the segments that are proximal and distal to the joint, the axis will be correct.

The figures for the normal ranges of joint motion have been revised in this edition. Upon reviewing a number of references, wide variations were found. For this reason, two middle figures were selected on the basis of a compilation of figures from the following sources: American Academy of Orthopaedic Surgeons, Esch and Epley, Hoppenfeld, Kapanji, Kendall and McCleary, Moore, and the Stanford Syllabi (see Reference list). A few ranges of motion were not included in all of the references.

A form for recording goniometric measurements similar to the ones for muscle examination grades is on page 11.

Notes Concerning the Text

In the accompanying anatomical information, taken largely from the American version of *Gray's Anatomy of the Human Body,* muscle origins and insertions on bones are given in some detail, but connective tissue attachments are included only when they are of particular significance.

In testing, the examiner should stand close to the patient in order that the manual force for stabilization or resistance can be given effectively and with minimal effort. It should be noted that in many of the illustrations, the examiner has stepped back from the table or is positioned on the side opposite to the one where the test ordinarily would be given in order that the full area could be drawn without obstruction or foreshortening.

Certain of the suggested testing positions should be modified for severely disabled patients. If it is necessary to use prone or sidelying positions instead of the sitting position, for example, it should be noted on the examination record.

The following sequence in testing is suggested to avoid too frequent turning of the patient, which may not only be fatiguing but may also result in increasing the length of time necessary for the tests:

TESTING POSITIONS

Supine Position

Neck
  Flexion—All tests

Trunk
  Flexion—All tests
  Rotation—All tests except Poor
  Elevation of Pelvis—All tests

Hip
  Flexion—Trace and Zero
  Flexion, Abduction and
    Lateral Rotation—Poor
                    Trace and Zero
  Abduction—Poor
            Trace and Zero
  Adduction—Poor
            Trace and Zero
  Lateral Rotation—Poor
                  Trace and Zero
  Medial Rotation—Poor
                  Trace and Zero

Knee
  Extension—Trace and Zero

Ankle and Foot
  Plantar Flexion—Normal and Good
                  Fair
  Dorsiflexion and Inversion—Trace and Zero
  Inversion—Poor
            Trace and Zero
  Eversion—Poor
           Trace and Zero

Toes (4 lateral) and Hallux—All tests

Scapula
  Abduction and Upward Rotation—
                          Normal and Good
                          Fair
  Elevation—Poor
            Trace and Zero

Shoulder
  Flexion—Trace and Zero
  Abduction—Poor
            Trace and Zero
  Horizontal Adduction—Normal and Good
                      Fair

Elbow
  Flexion—Poor
          Trace and Zero
  Extension—All tests

All tests for forearm, wrist, fingers and thumb can
be given in the supine position.

Prone Position

Neck
  Extension—All tests

Trunk
  Extension—All tests

Hip
  Extension—All tests except Poor

Knee
  Flexion—All tests except Poor

Scapula
  Adduction and Downward Rotation
    —Normal and Good
     Fair
  Adduction—Normal and Good
            Fair
  Elevation—Poor
            Trace and Zero
  Depression and Adduction—All tests

Shoulder
  Extension—All tests
  Horizontal Abduction—Normal and Good
                      Fair
  Lateral Rotation—All tests
  Medial Rotation—All tests

Sidelying Position

Hip
  Flexion—Poor
  Extension—Poor
  Abduction—Normal and Good
            Fair
  Abduction from
    Flexed Position—Normal and Good
                    Fair
  Adduction—Normal and Good
            Fair

Knee
  Flexion—Poor
  Extension—Poor

Ankle
  Plantar flexion—Poor
                  Trace and Zero

Foot
  Inversion—Normal and Good
            Fair
  Eversion—Normal and Good
           Fair

Sitting Position

Trunk
Rotation—Poor

Hip
Flexion—Normal and Good
        Fair
Flexion, Abduction and
   Lateral Rotation—Normal and Good
            Fair
Abduction from Flexed
   Position—Poor
        Trace and Zero
Lateral Rotation—Normal and Good
        Fair
Medial Rotation—Normal and Good
        Fair

Knee
Extension—Normal and Good
        Fair

Foot
Dorsiflexion and Inversion—Normal and Good
          Fair and Poor
Inversion—Fair

Scapula
Abduction and Upward Rotation—Poor
        Trace and Zero
Adduction and Downward Rotation—Poor
        Trace and Zero
Adduction—Poor
        Trace and Zero
Elevation—Normal and Good
        Fair

Shoulder
Flexion—Normal and Good
        Fair
        Poor
Abduction—Normal and Good
        Fair
Horizontal Abduction—Poor
        Trace and Zero
Horizontal Adduction—Poor
        Trace and Zero

Elbow
Flexion—Normal and Good
        Fair

All tests for forearm, wrist, fingers and thumb can
be given in the sitting position with forearm and
hand resting on a table.

Standing Position

Trunk
Elevation of Pelvis—Alternate Fair

Ankle
Plantar Flexion—Normal and Good
        Fair

MUSCLE EXAMINATION

Patient's Name _____ Chart No. _____

Date of Birth _____ Name of Institution _____

LEFT   RIGHT

			Examiner's Initials			
			Date			
			NECK Flexors Sternocleidomastoid			
			Extensor Group			
			TRUNK Flexors Rectus abdominis			
			R. Ext. abd. obl. / L. Int. abd. obl. } Rotators { L. Ext. abd. obl. / R. Int. abd. obl.			
			Extensors { Thoracic group / Lumbar group			
			Pelvic elev. Quadratus lumb.			
			HIP Flexors { Iliopsoas / Sartorius			
			Extensors Gluteus maximus			
			Abductors { Gluteus medius / Tensor fascia lata			
			Adductor group			
			Lateral rotator group			
			Medial rotator group			
			KNEE Flexors { Biceps femoris / Inner hamstrings			
			Extensors Quadriceps femoris			
			ANKLE Plantar flexors { Gastrocnemius / Soleus			
			FOOT Invertors { Tibialis anterior / Tibialis posterior			
			Evertors { Peroneus brevis / Peroneus longus			
			TOES M. P. flexors Lumbricals			
			I. P. flexors (Prox.) Flex. digit. br.			
			I. P. flexors (Distal) Flex. digit. l.			
			M. P. extensors { Ext. digit. l. / Ext. digit. br.			
			HALLUX M. P. flexor Flex. hall. br.			
			I. P. flexor Flex. hall. l.			
			M. P. extensor Ext. hall. br.			
			I. P. extensor Ext. hall. l.			
			GAIT:			

GRADING SYSTEM

Completes range of motion against gravity   Completes range of motion   No range of motion

N Normal — with full resistance at end of range   F Fair — against gravity   T Trace — slight contraction
G Good — with some resistance at end of range   P Poor — with gravity decreased   0 Zero — no contraction

9

MUSCLE EXAMINATION

LEFT							RIGHT			
				Examiner's Initials						
				Date						
			SCAPULA	Abductor	Serratus anterior					
				Elevator	Trapezius (superior)					
				Depressor	Trapezius (inferior)					
				Adductors	{ Trapezius middle) / Rhomboid maj. & min.					
			SHOULDER	Flexor	Deltoid (anterior)					
				Extensors	{ Latissimus dorsi / Teres major					
				Abductor	Deltoid (middle)					
				Horiz. abd.	Deltoid (posterior)					
				Horiz. add.	Pectoralis major					
				Lateral rotator group						
				Medial rotator group						
			ELBOW	Flexors	{ Biceps brachii / Brachialis / Brachioradialis					
				Extensor	Triceps brachii					
			FOREARM	Supinator group						
				Pronator group						
			WRIST	Flexors	{ Flex. carpi rad. / Flex. carpi uln.					
				Extensors	{ Ext. carpi rad. l. & br. / Ext. carpi uln.					
			FINGERS	M. P. flexors	Lumbricals					
				I. P. flexors (Prox.)	Flex. digit. sup.					
				I. P. flexors (Distal)	Flex. digit. prof.					
				M. P. extensor	Ext. digit. com.					
				Adductors	Palmar interossei					
				Abductors	Dorsal interossei					
				Abductor digiti minimi						
				Opponens digiti minimi						
			THUMB	M. P. flexor	Flex. poll. br.					
				I. P. flexor	Flex. poll. l.					
				M. P. extensor	Ext. poll. br.					
				I. P. extensor	Ext. poll. l.					
				Abductors	{ Abd. poll. br. / Abd. poll. l.					
				Adductor pollicis						
				Opponens pollicis						
			FACE:							

Additional data:

10

GONIOMETRIC MEASUREMENTS

Patient's Name _____ Chart No. _____

Date of Birth _____ Name of Institution _____

LEFT   RIGHT

				Examiner's Initials					
				Date					
				HIP	Flexion				
					Extension				
					Abduction				
					Adduction				
					Lateral Rotation				
					Medial Rotation				
				KNEE	Flexion				
					Extension				
				ANKLE	Plantar Flexion				
					Dorsiflexion				
				FOOT	Inversion				
					Eversion				
				TOES	Flexion				
					and				
					Extension				
				SHOULDER	Flexion				
					Extension				
					Abduction				
					Adduction				
					Lateral Rotation				
					Medial Rotation				
				ELBOW	Flexion				
					Extension				
				FOREARM	Supination				
					Pronation				
				WRIST	Flexion				
					Extension				
					Radial Deviation				
					Ulnar Deviation				
				FINGERS	Flexion-M.P.				
					Prox.				
					Dist.				
					Extension-M.P.				
					Prox.				
					Dist.				
					Abduction				
					Adduction				
				THUMB	Flexion-M.P.				
					Dist.				
					Extension-M.P.				
					Dist.				
					Abduction				
					Adduction				

FOR NOTES:

I INNERVATION CHARTS, MUSCLE TESTS, AND GONIOMETRIC MEASUREMENTS

Innervation of the Muscles of the Anterolateral Region of the Neck
(Showing derivation from the Cranial nerves and the Cervical plexus)

CRANIAL NERVES AND VENTRAL PRIMARY DIVISIONS OF SPINAL NERVES	Anterior Vertebral and Lateral Vertebral Muscles	Superficial and Lateral Cervical Muscles	Suprahyoid and Infrahyoid Muscles
Cr. 5 TRIGEMINAL		MANDIBULAR DIV.	INF. ALVEOLAR BR. MYLOHYOID NERVE Mylohyoid Digastric (ant. belly)
Cr. 7 FACIAL		FACIAL	Digastric (post. belly) Stylohyoid
		CERVICAL BRANCH Platysma	
Cr. 11 ACCESSORY			
Cr. 12 HYPOGLOSSAL			HYPOGLOSSI BRANCH (C1) through HYPOGLOSSAL Geniohyoid Thyrohyoid
C1		SPINAL ACCESSORY Trapezius (Cr. 11. C3. 4) Sternocleidomastoid (Cr. 11. C2. 3)	ANSA CERVICALIS (C1–3) Sternohyoid Sternothyroid Omohyoid
C2	LOOP (C1. 2) Rectus capitis anterior Rectus capitis lateralis (b)		
C3	(a) (b)		
C4	(a) (b)		
C5	*NERVE TO Levator Scapulae (a) (c)		
C6	*PHRENIC Diaphragm (a) (c)		
C7	(a) (c) (d)		
C8	(a) (c) (d)		
	(c) (d)		

(a) BRANCHES (C2–7)
  Longus colli
(b) BRANCHES (C1–3)
  Longus capitis

(c) BRANCHES (C6–8)
  Scalenus anterior
  Scalenus medius
(d) BRANCHES (C6–8)
  Scalenus posterior

(Nerves are in CAPITAL LETTERS)

*Although not innervating muscles of this region, they are included to complete muscular innervation from the cervical plexus.

14

Innervation of Muscles of the Posterior Region of the Neck and the Region of the Trunk

DORSAL PRIMARY DIVISIONS SPINAL NERVES

(C1) BRANCHES SUBOCCIPITAL

Suboccipital Muscles

(C4–8) LATERAL BRANCHES

Deep Muscles of the Neck and Back

Rectus capitis post. major
Rectus capitis post. minor
Obliquus capitis inferior
Obliquus capitis superior

BRANCHES SPINAL NERVES

Superficial Stratum
Splenius capitis
Splenius cervicis

Erector spinae
Iliocostalis
Longissimus
Spinalis

Deeper Stratum
Semispinalis capitis
Semispinalis cervicis
Multifidus
Rotatores
Interspinales
Intertransversarii medialis

VENTRAL PRIMARY DIVISIONS SPINAL NERVES

BRANCHES SPINAL NERVES

Intertransversarii
anteriores
posteriores
laterales

Muscles of the Thorax

Muscles of the Abdomen

(C3, 4, 5) PHRENIC

Diaphragm

(T1–11) INTERCOSTAL NERVES

Intercostales externi
Intercostales interni
Transversus thoracis
Levatores costarum

(T12) SUBCOSTAL NERVE

Subcostals

(T1–4) BRANCHES

Serratus posticus superior

(T9–12) BRANCHES

Serratus posticus inferior

(T7–12) BRANCHES INTERCOSTALS

(T12, L1) ILIOHYPOGASTRIC

(L1) ILIOINGUINAL

Anterior Muscles
External abdominal oblique
Internal abdominal oblique
Transversus abdominis
Rectus abdominis

(T12) BRANCH

Pyramidalis

(T12–L1) BRANCHES

Posterior muscles
Quadratus lumborum
Psoas major and minor*
Iliacus*

*See Innervation of the Muscles of the Lower Limb

(Nerves are in CAPITAL LETTERS)

15

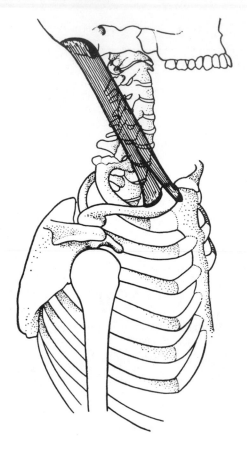

Lateral View
Sternocleidomastoid

Range of Motion:

Cervical spine flexes to just beyond point where convexity is straightened. (Greatest part of motion takes place in atlanto-occipital joint.)

Factors Limiting Motion:

1. Tension of posterior longitudinal ligament, ligamenta flava, and interspinal and supraspinal ligaments
2. Tension of posterior muscles of neck
3. Apposition of lower lips of vertebral bodies anteriorly with surfaces of subjacent vertebrae
4. Compression of intervertebral fibrocartilages in front

Fixation:

1. Contraction of anterior abdominal muscles
2. Weight of thorax and upper extremities

PRIME MOVER

MUSCLE	ORIGIN	INSERTION
Sternocleidomastoid N: Spinal accessory (Cr. 11) and ventral primary divisions (C2, 3)	*Sternal head:* a. Cranial part of ventral surface of manubrium sterni *Clavicular head:* a. Superior border and anterior surface of medial third of clavicle	a. Lateral surface of mastoid process from apex to superior border b. By thin aponeurosis into lateral half of superior nuchal line of occipital bone

Accessory Muscles

Longus capitis   Scalenus medius
Longus colli   Scalenus posterior
Scalenus anterior   Rectus capitis anterior
Infrahyoid group

NECK FLEXION

NORMAL AND GOOD

Supine, shoulders relaxed.

Stabilize lower thorax.

Patient flexes cervical spine through range of motion.

Resistance is given on forehead.

If movement is slow or appears difficult, the examiner should place one hand under the patient's head while the other gives resistance, as the muscles may "give way" suddenly.

Note: The Accessory muscles flex the head and stabilize the cervical spine as the Sternocleidomastoid muscles flex the neck.

If the Accessory muscles are weak, the contraction of strong Sternocleidomastoid muscles will increase rather than decrease the convexity of the cervical spine. The head can be raised but will be rotated posteriorly, chin up ("turtle neck position").

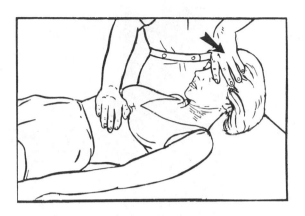

NORMAL AND GOOD

If there is a difference in strength of the two Sternocleidomastoid muscles, they may be tested separately by rotation of head to one side and flexion of neck.

Resistance is given above ear.

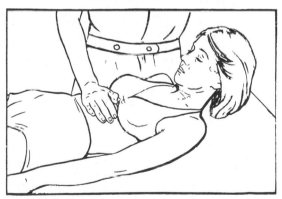

FAIR AND POOR

Supine, shoulders relaxed.

Stabilize lower thorax.

Patient flexes cervical spine through full range of motion for Fair grade and through partial range for Poor.

If flexion is difficult, examiner's hand should be placed under patient's head for protection.

Note: Test for Poor grade can be given with patient sidelying, examiner supporting head. However, there is a tendency to assist with the supporting hand.

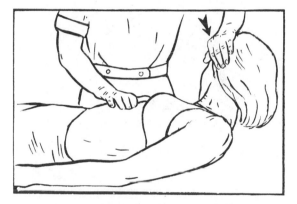

TRACE AND ZERO

The Sternocleidomastoid muscles may be palpated on each side of neck as patient attempts to flex.

Note: An attempted substitution by the Platysma would result in a pulling down of the corners of the mouth with a tendency to open the jaw and evidence of the contraction of subcutaneous muscle fibers over the anterior surface of the neck.

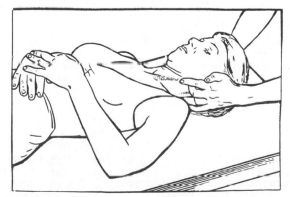

NECK EXTENSION

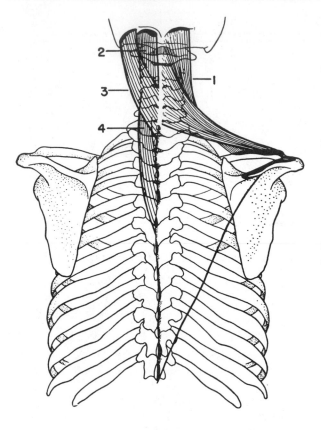

Range of Motion:

Cervical spine extends until head contacts dorsal muscle mass of upper trunk.

Factors Limiting Motion:

1. Tension of anterior longitudinal ligament of spine
2. Tension of ventral neck muscles
3. Approximation of spinous processes

Fixation:

1. Contraction of spinal extensor muscles of thorax and depressor muscles of scapulae and clavicles
2. Weight of trunk and upper extremities

Posterior View
1. *Trapezius (superior fibers)*
2. *Semispinalis capitis*
3. *Splenius capitis*
4. *Splenius cervicis*

PRIME MOVERS

Muscle	Origin	Insertion
Trapezius (superior fibers) N: Spinal Accessory (Cr. 11) and ventral primary divisions (C3, 4)	a. External occipital protuberance and medial third of superior nuchal line b. Upper part of ligamentum nuchae c. Spinous process of the seventh cervical	a. Dorsal border of lateral third of clavicle
Semispinalis capitis N: Dorsal primary divisions of the cervical nerves	a. Transverse processes of first 6 or 7 thoracic and seventh cervical vertebrae b. Articular processes of fourth, fifth, and sixth cervical vertebrae	a. Between superior and inferior nuchal lines of occipital bone
Splenius capitis N: Dorsal primary divisions (C4–8)	a. Caudal half of ligamentum nuchae b. Spinous processes of seventh cervical vertebra and first 3 or 4 thoracic vertebrae	a. Occipital bone just caudal to the lateral third of superior nuchal line b. Mastoid process of temporal bone

(Continued on Page 20)

NECK EXTENSION

NORMAL AND GOOD

Prone, with neck in flexion. Use pillow under thorax, if indicated.

Stabilize upper thoracic area and scapulae.

Patient extends cervical spine through range of motion.

Resistance is given on occiput.

Note: If there is considerable weakness of neck extensors or discomfort in this position, all tests may be given with head resting on table.

Extensor muscles on right may be tested by rotation of head to right with extension and vice versa.

See text and illustration at bottom of page concerning substitution of the back extensors.

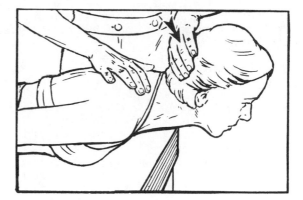

FAIR AND POOR

Prone, with neck flexed.

Stabilize upper thoracic area and scapulae.

Patient extends cervical spine through full range of motion for Fair grade or through partial range for Poor.

See Note under Neck Flexion (Poor grade).

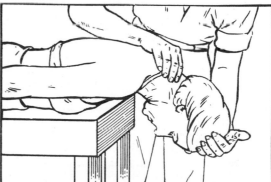

TRACE AND ZERO

Prone, head supported.

As patient attempts to extend, a trace may be determined by observation and palpation of the muscles of the dorsal area of the neck between C7 and occiput.

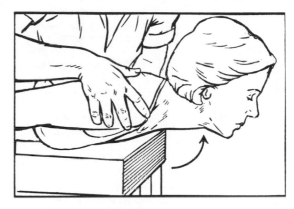

Substitution: Be sure patient completes full range of motion of neck extension. Back muscles may contract and lift upper trunk from table, giving the appearance of extension in cervical region.

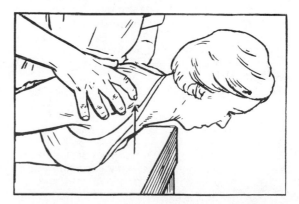

MUSCLE	ORIGIN	INSERTION
Splenius cervicis N: Dorsal primary divisions (C4–8)	a. Spinous processes of third to sixth thoracic vertebrae	a. Posterior tubercles of transverse processes of upper 2 or 3 cervical vertebrae
Erector spinae (Sacrospinalis) (Cervical and capitate sections not illustrated.) N: Dorsal primary divisions of adjacent spinal nerves		
Iliocostalis cervicis:	a. Angles of third to sixth ribs	a. Posterior tubercles of transverse processes of fourth, fifth, and sixth cervical vertebrae
Longissimus capitis:	a. Transverse processes of first 4 or 5 thoracic vertebrae b. Articular processes of last 3 or 4 cervical vertebrae	a. Posterior margin of mastoid process
Longissimus cervicis:	a. Transverse processes of first 4 or 5 thoracic vertebrae	a. Posterior tubercles of transverse processes of second to sixth cervical vertebrae
Spinalis capitis:	(Inseparably connected with Semispinalis capitis) a. Tips of transverse processes of first 6 or 7 thoracic and seventh cervical vertebrae b. Articular processes of last 3 cervical vertebrae	a. Between superior and inferior nuchal lines of occipital bone
Spinalis cervicis:	a. Caudal portion of ligamentum nuchae b. Spinous process of seventh cervical vertebra c. Sometimes from spinous processes of first and second thoracic vertebrae	a. Spinous process of axis b. Occasionally into spinous processes of second and third cervical vertebrae
Semispinalis cervicis N: Dorsal primary divisions of spinal nerves	a. Transverse processes of first 5 or 6 thoracic vertebrae	a. Spinous processes of axis to fifth cervical vertebrae

Accessory Muscles
Multifidus
Obliquus capitis superior and inferior
Rectus capitis posterior major and minor
Levator scapulae

TRUNK FLEXION

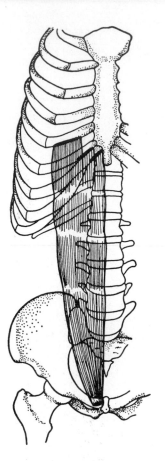

Anterior View
Rectus abdominis

Range of Motion:

In supine position, flexion of thorax on pelvis is possible until scapulae are raised from table. Motion takes place primarily in thoracic spine. (Trunk is carried to sitting position by reverse action of hip flexors with abdominal muscles acting as fixators.)

Factors Limiting Motion:

1. Tension of posterior longitudinal ligament, ligamenta flava, and interspinal and supraspinal ligaments
2. Tension of spinal extensor muscles
3. Apposition of caudal lips of vertebral bodies anteriorly with surfaces of subjacent vertebrae
4. Compression of ventral part of intervertebral fibrocartilages
5. Contact of last ribs with abdomen

Fixation:

1. Reverse action of hip flexor muscles
2. Weight of legs and pelvis

PRIME MOVER

Muscle	Origin	Insertion
Rectus abdominis N: Intercostal nerves (7–12)	a. Crest of pubis b. Ligaments covering ventral surface of symphysis pubis	a. By 3 portions into cartilages of fifth, sixth, and seventh ribs

Accessory Muscles
Internal abdominal oblique
External abdominal oblique (reverse action)

TRUNK FLEXION

NORMAL

Supine with hands behind neck.

Stabilize lower limbs.

Patient flexes trunk through range of motion. A "curl-up" is emphasized, and flexion is possible until scapulae are raised from table.

If abdominals are weak, the reverse action of the hip flexors may cause lumbar lordosis. If so, the patient's hips and knees should be flexed (feet flat on table). This allows "slack" in the hip flexors. However, if the extensor muscles of the lumbar spine are weak, contraction of the abdominal muscles can cause posterior tilting of the pelvis. If this occurs, tension in the hip flexors would be useful in stabilizing the pelvis.

If hip flexors are weak, stabilize the pelvis.

Note: In all tests, observe the umbilicus. Cranialward movement indicates a stronger contraction of upper segments of muscle, and caudialward movement indicates a stronger contraction of lower segments.

Tests for neck flexion should precede those for trunk flexion.

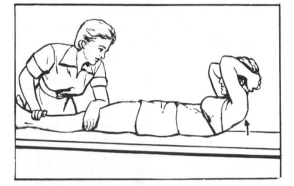

GOOD

Supine with arms at sides.

Stabilize lower limbs.

Patient flexes through range of motion.

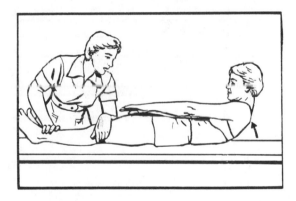

FAIR

Supine with arms at sides.

Stabilize lower limbs.

Patient flexes trunk through partial range of motion. Head, tips of shoulders and cranial borders of scapulae should clear table with inferior angle remaining in contact with table.

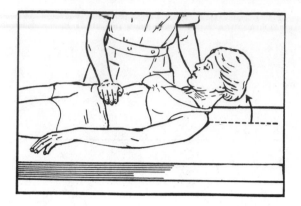

TRUNK FLEXION

POOR

Supine with arms at sides.

Patient flexes cervical spine. Caudal portion of thorax is depressed, and pelvis is tilted until the lumbar area of spine is flat on table.

Palpation will help to determine smoothness of contraction.

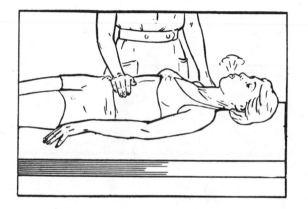

TRACE AND ZERO

Supine.

A slight contraction may be determined by palpation over anterior abdominal wall as patient attempts to cough (also during rapid exhalation or as patient attempts to lift head).

TRUNK ROTATION

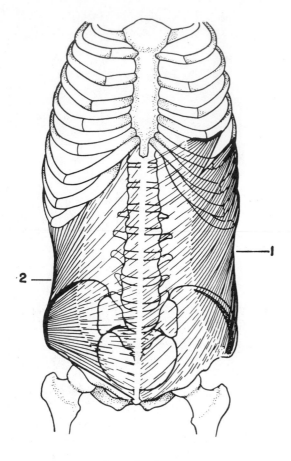

Anterior View

Range of Motion

In supine position, rotation of thorax is possible until scapula on side of forward shoulder is raised from table.

Factors Limiting Motion:

1. Tension of anulus fibrosus between vertebrae
2. Tension of oblique abdominal muscles on side opposite those being tested
3. In thoracic area, tension of costovertebral ligaments
4. In lumbar area, interlocking of articular facets. (Rotation negligible.)

Fixation:

Reverse action of hip flexor muscles.

1. *External abdominal oblique*
2. *Internal abdominal oblique*

PRIME MOVERS

MUSCLE	ORIGIN	INSERTION
External abdominal oblique N: Intercostals (T8–12) Iliohypogastric (T12, L1) Ilioinguinal (L1)	a. Eight digitations from external surfaces and inferior borders of lower 8 ribs	a. Anterior half of outer lip of iliac crest b. Aponeurosis to pubic tubercle and pectineal line in middle, interlaces with aponeurosis of opposite muscle forming linea alba, extending from xiphoid process to symphysis pubis
Internal abdominal oblique N: Intercostals (T8–12) Iliohypogastric (T12, L1) and sometimes Ilioinguinal (L1)	a. Lateral half of inguinal ligament b. Anterior two-thirds of middle lip of iliac crest c. Posterior layer of the thoracolumbar fascia near the crest	a. Crest of pubis and medial part of pectineal line b. Linea alba c. Cartilages of seventh, eighth, and ninth ribs d. Inferior borders of cartilages of the last 3 or 4 ribs

Accessory Muscles

Latissimus dorsi Semispinalis	Multifidus Rotatores	Rectus abdominis (combined trunk rotation and flexion)

TRUNK ROTATION

NORMAL

Supine with hands behind neck.

Stabilize lower limbs.

Patient flexes trunk and rotates thorax to one side. Repeat to opposite side.

Test for left External abdominal oblique and right Internal abdominal oblique is shown in illustration. Rotation to left is brought about by opposite muscles.

Note: In all tests, observe any deviation of umbilicus, which will move toward strongest quadrant if there is a difference in strength of opposing oblique muscles.

Flaring of rib cage denotes weakness of external obliques.

If hip flexor muscles are weak, stabilize pelvis.

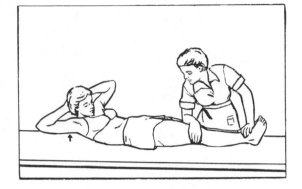

GOOD AND FAIR

Supine with arms at sides.

Stabilize lower limbs.

Patient flexes trunk and rotates thorax to one side.

A Good grade is given if scapula on forward shoulder is raised from table and other partially raised.

For a Fair grade, the scapula on the forward shoulder only is raised from table.

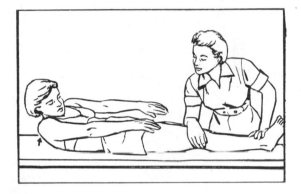

POOR

Sitting with arms relaxed at sides.

Stabilize pelvis.

Patient rotates thorax. Repeat with rotation to opposite side.

If sitting position is contraindicated, a partial range of motion in the test for Fair may be graded Poor.

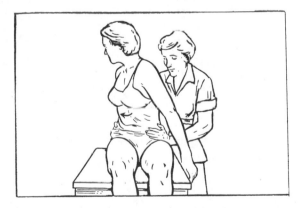

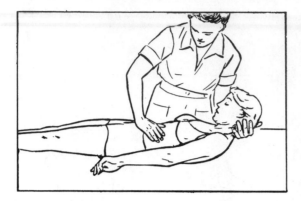

TRUNK ROTATION

TRACE AND ZERO

Examiner palpates muscles below the lower edge of the ribs as patient attempts to approximate thorax on left and pelvis on right. Repeat on opposite side.

Note: If an attempt is made to substitute the Pectoralis major, the shoulder will be raised from table with limited rotation of the trunk.

TRUNK EXTENSION

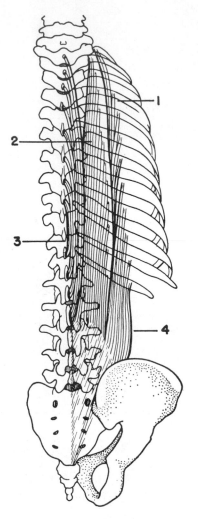

Posterior View

Range of Motion:

    Thoracic spine extends only to approximately a straight line. Lumbar spine extends freely.

Factors Limiting Motion:

    1. Tension of anterior longitudinal ligament of spine
    2. Tension of anterior abdominal muscles
    3. Contact of spinous processes
    4. Contact of caudal articular margins with laminae

Fixation:

    1. Contraction of Gluteus maximus and hamstring muscles
    2. Weight of pelvis and legs

Erector spinae (Sacrospinalis)

1. *Iliocostalis thoracis*
2. *Longissimus thoracis*
3. *Spinalis thoracis*
4. *Iliocostalis lumborum*

PRIME MOVERS

Muscle	Origin	Insertion
Erector spinae (Sacrospinalis) N: Dorsal primary divisions adjacent spinal nerves		
Iliocostalis thoracis	a. Upper borders of angles of the last 6 ribs medial to Iliocostalis lumborum	a. Cranial borders of angles of first 6 ribs b. Transverse process of seventh cervical vertebra
Longissimus thoracis	a. Common tendon of Erector spinae b. Transverse processes of lumbar vertebrae c. Anterior layer of lumbocostal aponeurosis	a. Tips of transverse processes of all thoracic vertebrae b. Last 9 or 10 ribs between their tubercles and angles

(Continued on Page 32)

TRUNK EXTENSION

NORMAL AND GOOD

(Extension of lumbar spine)

Prone with a pillow under the abdomen for the comfort of the patient and to gain a greater range of motion. (Latter not illustrated.)

Stabilize pelvis.

With arms and shoulders clear of table to prevent their use in extending trunk, patient extends lumbar spine until caudal part of thorax is raised from table.

Resistance is given on caudal portion of thoracic area.

Note: Tests for neck extension should precede those for trunk extension.

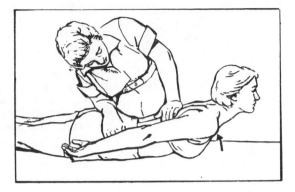

NORMAL AND GOOD

(Extension of thoracic spine)

Prone with a pillow under the abdomen.

Stabilize pelvis and lower part of thorax.

Patient extends thoracic spine to horizontal position.

Resistance is given on cranial portion of thorax.

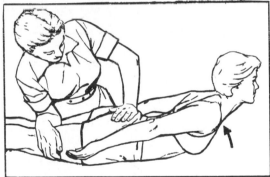

FAIR

(Extension of thoracic and lumbar spine)

Prone.

Stabilize pelvis and lower limbs.

Patient extends thoracic and lumbar spine through range of motion.

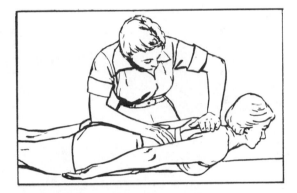

POOR

(Extension of thoracic and lumbar spine)

Prone.

Stabilize pelvis and lower limbs.

Patient completes partial range of motion. (Not illustrated.)

TRACE AND ZERO

Prone.

Examiner palpates spinal extensor muscles to determine presence and degree of contraction as patient attempts to extend spine.

MUSCLE	ORIGIN	INSERTION
Spinalis thoracis	a. Spinous processes of first 2 lumbar and last 2 thoracic vertebrae	a. Spinous processes of first 4 to 8 thoracic vertebrae
Iliocostalis lumborum	a. Common tendon: middle crest of sacrum, spinous processes of lumbar and eleventh and twelfth thoracic vertebrae, supraspinal ligament, posterior portion of inner lip of iliac crest, and lateral crest of sacrum	a. Inferior borders of angles of last 6 or 7 ribs
Quadratus lumborum N: Ventral primary divisions (T12, L1)	a. Iliolumbar ligament and adjacent 5 cm of iliac crest	a. Medial half of inferior border of last rib b. Apices of transverse processes of first 4 lumbar vertebrae

Accessory Muscles

Semispinalis   Rotatores
Multifidus

ELEVATION OF PELVIS

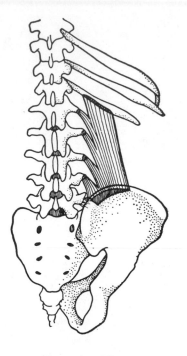

Posterior View
Quadratus lumborum

Range of Motion:

In standing position pelvis may be raised on one side until foot is well clear of floor. (Reverse action of Quadratus lumborum.)

Factors Limiting Motion:

1. Tension of spinal ligaments on opposite side
2. Contact of iliac crest with thorax

Fixation:

Contraction of spinal extensor muscles to fix thorax

PRIME MOVERS

Muscle	Origin	Insertion
Quadratus lumborum (reverse action) N: Ventral primary divisions (T12, L1)	a. Iliolumbar ligament and adjacent 5 cm of iliac crest	a. Medial half of inferior border of last rib b. Small tendons into apices of transverse processes of first 4 lumbar vertebrae
Additional portion occasionally present:	a. Transverse processes of last 3 or 4 lumbar vertebrae	a. Inferior margin of last rib
Iliocostalis lumborum (reverse action; illustrated on p. 30) N: Dorsal primary divisions of adjacent spinal nerves	a. Common tendon: middle crest of sacrum, spinous processes of lumbar and eleventh and twelfth thoracic vertebrae, supraspinal ligament, posterior portion of inner lip of iliac crest, and lateral crest of sacrum	a. Inferior borders of angles of last 6 or 7 ribs

Accessory Muscles

External abdominal oblique (lateral fibers)
Internal abdominal oblique (lateral fibers; reverse action)

Latissimus dorsi (with arms fixed, reverse action)
Hip abductor muscles (on opposite side; reverse action)

ELEVATION OF PELVIS

NORMAL AND GOOD

Supine (or prone) with lumbar area of spine in moderate extension. Patient grasps edge of table to stabilize thorax. If arm and shoulder muscles are weak, an assistant should stabilize thorax.

Patient draws pelvis toward thorax on one side.

Resistance is given above ankle joint.

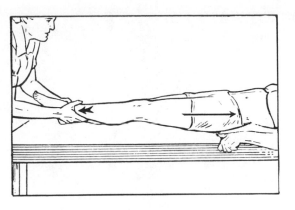

FAIR AND POOR

Supine with lumbar area of spine in moderate extension.

Patient may grasp side of table to stabilize thorax (not shown in drawing.)

Patient draws pelvis toward thorax.

Slight resistance is given for a Fair grade.

Completion of range is graded Poor.

Note: Patient may attempt substitution by lateral flexion of trunk on the contralateral side.

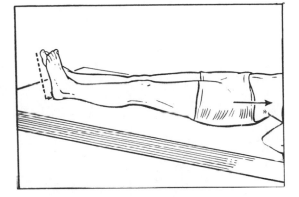

FAIR

(Alternate)

Standing position.

Stabilize thorax.

Patient lifts pelvis toward thorax through range of motion.

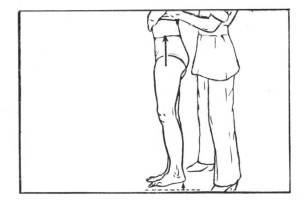

TRACE AND ZERO

A contraction of the Quadratus lumborum may be determined by deep palpation in lumbar area under lateral edge of Erector spinae as patient attempts to draw pelvis toward thorax.

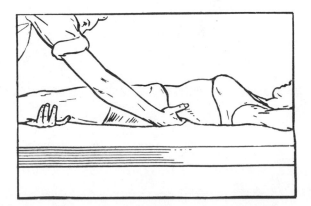

FOR NOTES:

INNERVATION OF THE MUSCLES OF THE LOWER LIMB
(Showing Derivation from the Lumbosacral Plexus)

VENTRAL PRIMARY DIVISIONS OF SPINAL NERVES

Iliac Region and Thigh

Leg and Foot

L2
L3
L4
L5
S1
S2
S3

BRANCHES TO
Psoas major

DORSAL DIV. (L2, 3)
LATERAL FEMORAL CUTANEOUS

ANT. (L2, 3, 4)
- Adductor longus
- Gracilis
- Adductor brevis

VENTRAL DIV. (L2, 3, 4) OBTURATOR

POST. (L3, 4)
- Adductor magnus
- Obturator externus

DORSAL DIV. (L2, 3, 4) FEMORAL
- Iliacus (L2, 3)
- Sartorius (L2, 3)
- Quadriceps femoris
- Rectus femoris
- Vastus intermedius
- Vastus medialis
- Vastus lateralis
- Articularis genu

VENTRAL DIV. (L3, 4)
ACCESSORY OBTURATOR
- Pectineus (L2, 3, 4)

LUMBOSACRAL TRUNK (L4, 5)

SCIATIC (L4, 5, S1, 2, 3)

TIBIAL (L4, 5, S1, 2, 3)
- Biceps femoris (S1, 2, 3) (long head)
- Semitendinosus (L5, S1, 2)
- Semimembranosus (L5, S1)
- Adductor magnus (L3, 4)

COMMON PERONEAL (L4, 5, S1, 2)
- Biceps femoris (L5, S1, 2) (short head)

(d) (e)

(a) (b) (c)

TIBIAL
- Gastrocnemius (S1, 2)
- Soleus (S1, 2)
- Tibialis posterior (L5, S1)
- Plantaris (L4, 5, S1)
- Popliteus (L4, 5, S1)
- Semimembranosus (L5, S1)
- Adductor magnus (L3, 4)

MEDIAL PLANTAR (L4, 5, S1)
- Flexor digitorum brevis (L4, 5)
- Abductor hallucis (L4, 5)
- 1st lumbrical (L4, 5)
- Flexor hallucis brevis (L4, 5, S1)

LATERAL PLANTAR (S1, 2)
- Abductor digiti minimi
- Quadratus plantae
- Adductor hallucis
- 2nd, 3rd and 4th lumbricals
- Plantar interossei
- Dorsal interossei

SUPERFICIAL PERONEAL (L4, 5, S1)
- Peroneus longus
- Peroneus brevis

DEEP PERONEAL (L4, 5, S1)
- Tibialis anterior
- Extensor digitorum longus
- Peroneus tertius
- Extensor hallucis longus

DEEP PERONEAL (L5, S1)
- Extensor digitorum brevis

(a) INFERIOR GLUTEAL (L5, S1, 2)
  Gluteus maximus
(b) NERVE TO OBTURATOR INTERNUS AND GEMELLUS SUP. (L5, S1, 2)
(c) BRANCHES TO PIRIFORMIS (S1, 2)
(d) NERVE TO QUADRATUS FEMORIS AND GEMELLUS INF. (L4, 5, S1)
(e) SUPERIOR GLUTEAL (L4, 5, S1)
  Gluteus medius
  Gluteus minimus
  Tensor fascia lata

(Nerves are in CAPITAL LETTERS)

—Chart adapted from Worthingham, C. *Upper and Lower Extremity Muscle and Innervation Charts.* Stanford University Press, 1944.

37

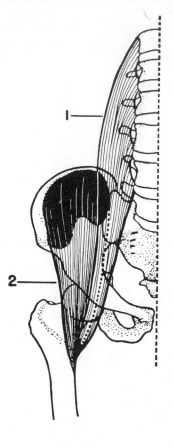

Anterior
1. *Psoas Major*
2. *Iliacus*

Range of Motion:

0 to 120–130 degrees

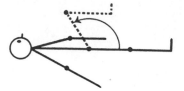

Factor Limiting Motion:

Contact of thigh with pelvis

Measurement:

Supine, hip and knee are flexed. Contralateral thigh is pressed strongly against table by patient to lessen posterior tilting of pelvis.

1. Place stationary arm of the goniometer on lateral longitudinal midline of trunk.
2. Movable arm is placed on lateral longitudinal midline of thigh.

Reading is taken as thigh contacts pelvis. Any additional movement toward the knee-chest position is brought about by a posterior tilting of pelvis and flexion in lumbar area of spine.

PRIME MOVERS

Muscle	Origin	Insertion
Psoas major N: Ventral primary divisions (L2, 3)	a. Transverse processes of all lumbar vertebrae b. Sides of bodies of last thoracic and all lumbar vertebrae and corresponding ventral surfaces of the intervertebral disks	a. Lesser trochanter of femur
Iliacus N: Femoral (L2, 3)	a. Superior two thirds of iliac fossa b. Inner lip of iliac crest c. Base of sacrum	a. Lateral side of tendon of psoas major b. Body of femur distal to lesser trochanter

Accessory Muscles

Rectus femoris   Pectineus
Sartorius   Adductor brevis
Tensor fascia lata   Adductor longus
Adductor magnus (oblique fibers)

HIP FLEXION

NORMAL AND GOOD

Sitting with legs over edge of table. Patient grasps table edge to help stabilize trunk.

Stabilize pelvis in a posterior tilt. (Examiner should stand close to patient and use her body weight carefully in giving resistance. In illustration, she has stepped back to offer a clear view of the placement of her hands.)

Patient flexes hip through last part of range of motion.

Resistance is given proximal to knee joint.

Note: If the sitting position is contraindicated, tests may be carried out with the patient supine. For a Fair grade, slight resistance should be given at the end of the range of motion as gravity assists motion after 90 degrees of flexion.

See possible substitutions on next page.

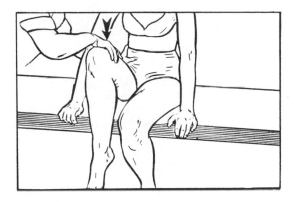

FAIR

Sitting with legs over edge of table, and hands grasping table edge.

Stabilize pelvis in a posterior tilt.

Patient flexes hip through last part of range of motion.

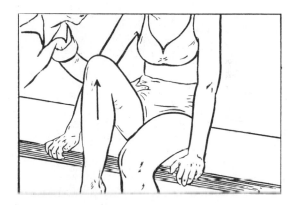

POOR

Sidelying with upper limb supported. Trunk and limbs straight.

Stabilize pelvis in a posterior tilt.

Patient flexes hip through range of motion. Knee is allowed to flex to prevent hamstring tension.

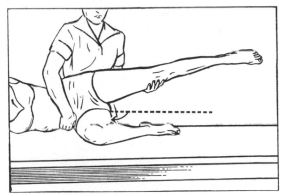

TRACE AND ZERO

Supine with limb supported. It may be possible to detect contraction in Psoas major just distal to inguinal ligament on medial side of Sartorius as patient attempts to flex hip.

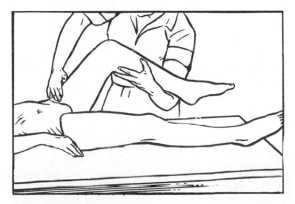

HIP FLEXION

Substitution by Sartorius will cause lateral rotation and abduction of hip. Muscle may easily be seen and palpated near its origin during the motion.

Note: Substitution by Tensor fascia lata causes medial rotation and abduction of the hip. Muscle may be seen and palpated at its origin.

Patient may lean backward in an attempt to substitute accessory muscles. The prime movers are mainly responsible for the last part of the range of motion.

HIP FLEXION, ABDUCTION AND LATERAL ROTATION WITH KNEE FLEXION

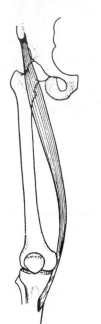

Anterior View
Sartorius

Range of Motion:

Combined joint action; ranges of motion incomplete

Factors Limiting Motion:

None; ranges of motion incomplete

Fixation:

1. Contraction of abdominal muscles to fix pelvis
2. Weight of trunk

PRIME MOVER

MUSCLE	ORIGIN	INSERTION
Sartorius N: Femoral (L2, 3)	a. Anterior superior spine of ilium b. Upper half of notch on anterior border of bone immediately distal to spine	a. Anteromedial surface of tibial shaft posterior to tuberosity (anterior to Gracilis and Semitendinosus)
	Accessory Muscles Hip and knee flexors Hip external rotators Hip abductors	

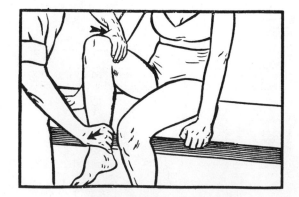

HIP FLEXION, ABDUCTION AND LATERAL ROTATION WITH KNEE FLEXION

NORMAL AND GOOD

Sitting with legs over side of table.

Patient flexes, abducts and laterally rotates hip and flexes knee.

Resistance to hip flexion and abduction is given with one hand above knee joint; resistance to hip lateral rotation and knee flexion is given with other hand above ankle joint.

Note: If the sitting position is contraindicated, all tests can be given supine, with slight resistance for a Fair grade.

HIP FLEXION, ABDUCTION AND LATERAL ROTATION WITH KNEE FLEXION

FAIR

Sitting with legs over side of table; heel of limb to be tested in front of opposite ankle.

Stabilize pelvis in a posterior tilt.

Patient raises heel to knee with flexion, abduction and lateral rotation of hip and flexion of knee.

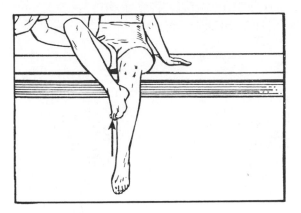

POOR

Supine with heel of limb to be tested on opposite ankle.

Stabilize pelvis.

Patient slides heel along leg to knee with flexion, abduction and lateral rotation of hip and flexion of knee.

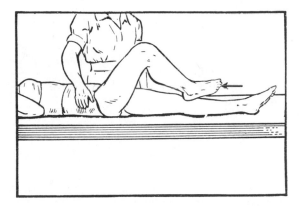

TRACE AND ZERO

Tendinous fibers of Sartorius may be found near origin just below anterior superior spine of ilium.

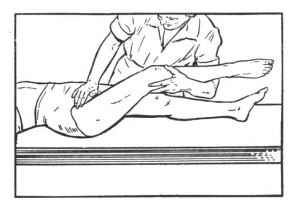

Substitution of Iliopsoas or Rectus femoris in this movement is evidenced by straight hip flexion without abduction and lateral rotation.

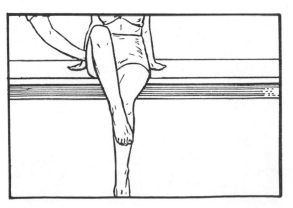

HIP EXTENSION

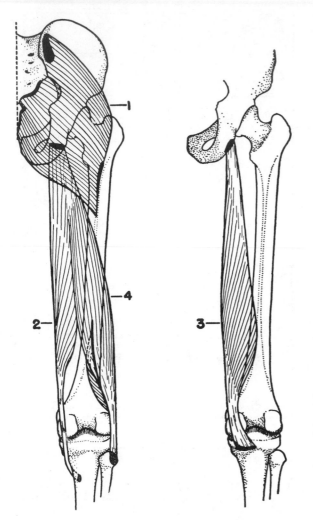

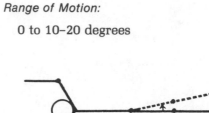

Range of Motion:

0 to 10–20 degrees

EXTENSION BEYOND MIDLINE

Factors Limiting Motion:

1. Tension of iliofemoral ligament
2. Tension of hip flexor muscles

Measurement:

Prone, both hips and knees are straight. Contralateral thigh is pressed strongly against table by patient to lessen anterior tilting of the pelvis as hip to be measured is extended.

1. Place stationary arm of goniometer on lateral longitudinal midline of trunk.
2. Movable arm is placed on lateral longitudinal midline of thigh.

Do not allow pelvis to be lifted from the table.

Posterior View

1. *Gluteus maximus*
2. *Semitendinosus*
3. *Semimembranosus*
4. *Biceps femoris (long head)*

PRIME MOVERS

MUSCLE	ORIGIN	INSERTION
Gluteus maximus N: Inferior gluteal (L5, S1, 2)	a. Posterior gluteal line and lateral lip of iliac crest superior and dorsal to line b. Posterior surface of lower part of sacrum and side of coccyx c. Posterior surface of sacrotuberous ligament and aponeurosis of the Erector spinae	a. Iliotibial band of fascia lata over greater trochanter b. Gluteal tuberosity
Semitendinosus N: Sciatic (tibial portion) (L5, S1, 2, 3)	a. Distal and medial impression on ischial tuberosity	a. Anteromedial surface of tibia at proximal end of shaft

44

(Continued on Page 46)

HIP EXTENSION

NORMAL AND GOOD

Prone.
Stabilize pelvis. (Examiner should stand close to patient on same side of table as extremity being tested.)
Patient extends hip through range of motion.
Resistance is given proximal to knee joint.

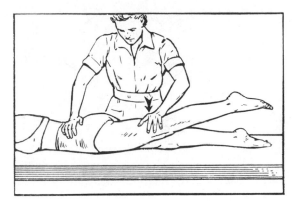

NORMAL AND GOOD

(Test for isolation of Gluteus maximus)

Prone with knee flexed.
Stabilize pelvis.
Patient extends hip, keeping knee flexed to decrease action of hamstrings.
Resistance is given proximal to knee joint.
Range of motion will be more limited than in position above, owing to tension in the Rectus femoris.
Note: Knee should be flexed less than 90 degrees if muscles tend to cramp.

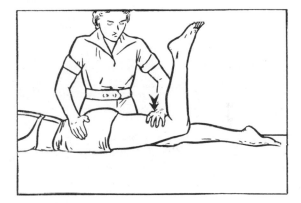

FAIR

Prone.
Stabilize pelvis.
Patient extends hip through range of motion.

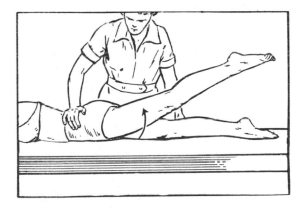

POOR

Sidelying with hip flexed, knee extended and upper limb supported.
Stabilize pelvis.
Patient extends hip through full range of motion. Knee may be flexed for Fair and Poor to isolate the action of the Gluteus maximus.

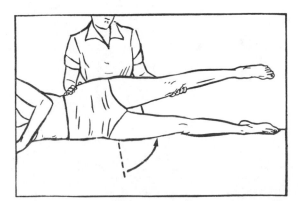

TRACE AND ZERO

Prone.

Contraction of Gluteus maximus will result in narrowing of gluteal crease. Lower and upper sections of muscle should be palpated as patient attempts to extend hip.

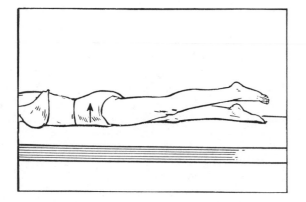

Patient may lift pelvis and support limb with hamstrings, raising limb from table by extending lumbar spine. Examiner must be certain that pelvis is stable and movement takes place in hip joint.

Note: If there is a contracture of the hip flexors, or a greater range of motion is desired, test may be given with patient prone, hips flexed to 90 degrees over the end of the table. (Not recommended for elderly patients or those with generalized weakness who may be apprehensive in this position. Make an estimate with patient sidelying.)

PRIME MOVERS *(Continued)*

MUSCLE	ORIGIN	INSERTION
Semimembranosus N: Sciatic (tibial portion) (L5, S1, 2)	a. Proximal and lateral impression on ischial tuberosity	a. Horizontal groove on posterior medial aspect of medial condyle of tibia b. Fibrous expansion into fascia covering Popliteus, the tibial collateral ligament, and the fascia of the leg
Biceps femoris (long head) N: Sciatic (tibial portion) (S1, 2, 3)	*Long head:* a. Distal and medial impression on ischial tuberosity	a. Lateral side of head of fibula b. Slip to lateral condyle of tibia

HIP ABDUCTION

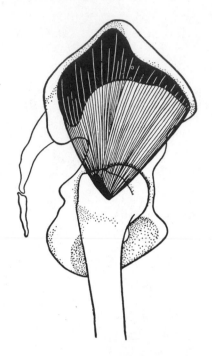

Lateral View
Gluteus medius

Range of Motion:

0 to 45 degrees

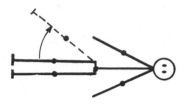

Factors Limiting Motion:

1. Tension of distal band of iliofemoral ligament and pubocapsular ligament
2. Tension of hip adductor muscles

Measurement:

Supine, knee extended and hip abducted.

1. Place stationary arm of the goniometer parallel to but below both anterior superior spines of the ilium at the level of hip joint.
2. Movable arm is placed on the anterior longitudinal midline of thigh.

Do not allow lateral hip rotation or movement of pelvis toward lower ribs.

PRIME MOVER

MUSCLE	ORIGIN	INSERTION
Gluteus medius N: Superior gluteal (L4, 5, S1)	a. Outer surface of ilium between iliac crest and posterior gluteal line dorsally and anterior gluteal line ventrally b. Gluteal aponeurosis	a. Oblique ridge on lateral surface of greater trochanter

Accessory Muscles
Gluteus minimus
Tensor fascia lata
Gluteus maximus (upper fibers)

Hip Abduction

NORMAL AND GOOD

Sidelying with hip slightly extended beyond midline. Lower knee flexed for balance.

Stabilize pelvis.

Patient abducts hip through range of motion without lateral rotation.

Resistance is given proximal to knee joint.

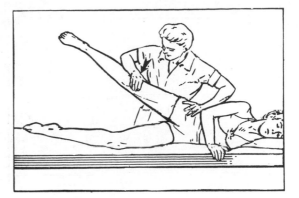

FAIR

Sidelying with hip slightly extended beyond midline. Lower knee flexed for balance.

Stabilize pelvis.

Patient abducts hip through range of motion.

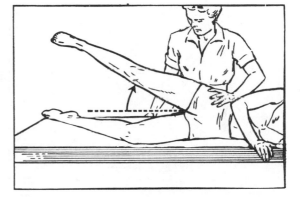

POOR

Supine.

Stabilize pelvis and contralateral limb.

Patient abducts hip through range of motion without rotation.

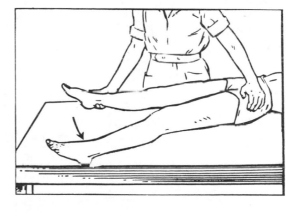

TRACE AND ZERO

Fibers of the Gluteus medius may be palpated on lateral aspect of ilium proximal to the greater trochanter of femur as patient attempts to abduct hip.

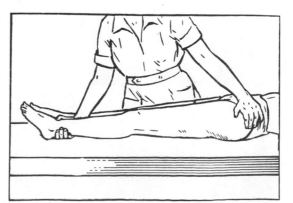

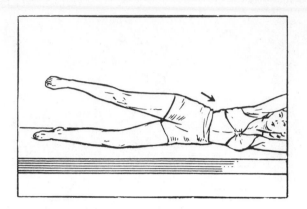

Patient may bring pelvis to thorax by strong contraction of lateral trunk muscles, thereby lifting limb through partial abduction. Examiner must stabilize pelvis to make sure motion takes place in hip joint. (Arm is pictured overhead to permit view of contracting trunk muscles.)

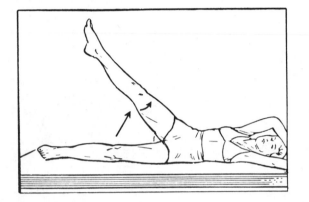

Lateral rotation at the hip should be eliminated during hip abduction or hip flexors may substitute for Gluteus medius. Flexion of the hip allows substitution by the Tensor fascia lata.

HIP ABDUCTION FROM FLEXED POSITION

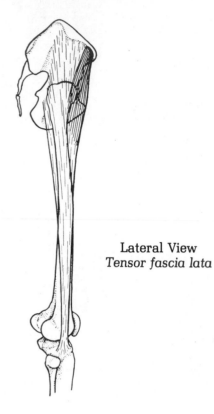

Lateral View
Tensor fascia lata

Range of Motion:

Combined joint action; ranges of motion incomplete

Factors Limiting Motion:

None; ranges of motion incomplete

Fixation:

1. Contraction of lateral abdominal muscles and Latissimus dorsi
2. Weight of trunk

PRIME MOVER

MUSCLE	ORIGIN	INSERTION
Tensor fascia lata N: Superior gluteal (L4, 5, S1)	a. Anterior part of outer lip of iliac crest b. Outer surface of anterior superior iliac spine c. Deep surface of the fascia lata.	a. Between 2 layers of the iliotibial band at juncture of middle and proximal thirds (iliotibial band inserts on lateral condyle of tibia)

Accessory Muscles
Gluteus medius
Gluteus minimus

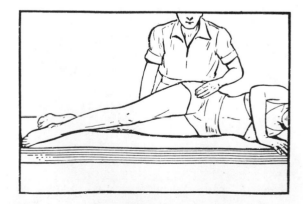

HIP ABDUCTION FROM FLEXED POSITION

NORMAL AND GOOD

Sidelying with lower knee slightly flexed for balance; limb to be tested flexed to angle of approximately 45 degrees at hip joint.
Stabilize pelvis.

HIP ABDUCTION FROM FLEXED POSITION

NORMAL AND GOOD

(Continued)

Patient abducts hip through range of motion of approximately 30 degrees.
Resistance is given above knee joint.

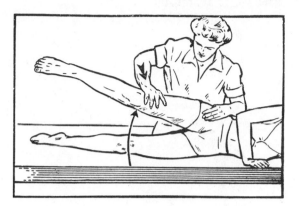

FAIR

Sidelying with lower knee slightly flexed for balance; limb to be tested flexed to 45 degrees at hip joint.
Stabilize pelvis.
Patient abducts hip through range of motion of approximately 30 degrees.

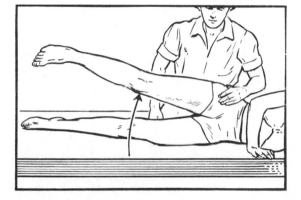

POOR

Sitting on table with knees extended. Trunk at a 45 degree angle to table and supported by patient's arms behind back.
Stabilize pelvis.
Patient abducts hip through range of motion of approximately 30 degrees.

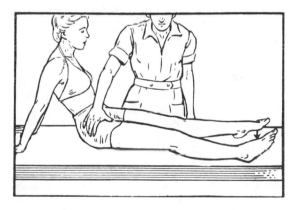

TRACE AND ZERO

Observe and palpate below origin of muscle and at fascial insertion on lateral side of knee joint to determine presence of contraction in Tensor.

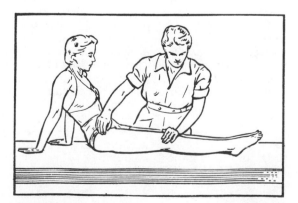

HIP ADDUCTION

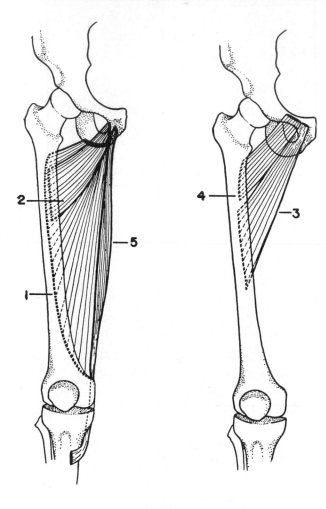

Range of Motion:

0 to 20–30 degrees

Factors Limiting Motion:

1. Contact with opposite limb
2. When hip is flexed, tension of ischio-femoral ligament

Measurement:

Supine, uninvolved hip abducted. Limb to be measured adducted at the hip as far past zero position as possible.

1. Place stationary arm of goniometer parallel to but below both anterior superior spines of the ilium at level of hip joint.
2. Movable arm is placed on the anterior longitudinal midline of thigh.

Do not allow medial rotation of the hip.

Anterior View
1. *Adductor magnus*
2. *Adductor brevis*
3. *Adductor longus*
4. *Pectineus*
5. *Gracilis*

PRIME MOVERS

Muscle	Origin	Insertion
Adductor magnus N: Obturator (posterior division L3, 4), branch from sciatic (common peroneal division L3, 4)	a. Outer margin of inferior surface of tuberosity of ischium b. Inferior ramus of ischium c. Anterior surface of inferior ramus of pubis	a. Broad aponeurosis into whole length of linea aspera and its medial prolongation b. Tendon into adductor tubercle on medial condyle of femur
Adductor brevis N: Obturator (anterior division L3, 4)	a. Outer surface of inferior ramus of pubis	a. Aponeurosis, distal two thirds of line leading from lesser trochanter to linea aspera, and into the proximal part of linea aspera

(Continued on Page 56)

HIP ADDUCTION

NORMAL AND GOOD

Sidelying with limb to be tested resting on table and upper one supported in approximately 25 degrees of abduction.

Patient adducts hip until thigh contacts upper one.

Resistance is given proximal to knee joint with counterpressure against upper limb.

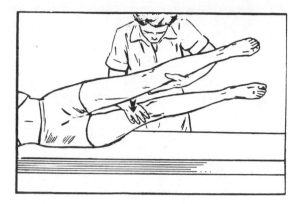

FAIR

Sidelying with limb to be tested resting on table and upper limb supported in approximately 25 degrees of abduction.

Patient adducts hip until thighs are in contact.

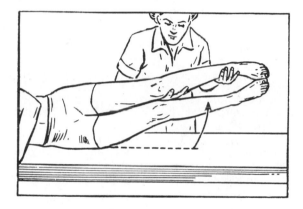

POOR

Supine, contralateral hip abducted to approximately 25 degrees.

Stabilize pelvis and contralateral limb.

Patient adducts hip through range of motion without rotation.

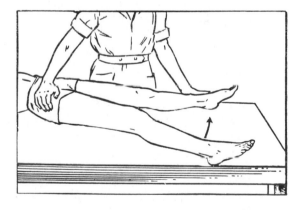

TRACE AND ZERO

Contraction of fibers of adductor muscles may be palpated on medial aspect of thigh and the proximal attachments near the pubic rami as hip adduction is attempted.

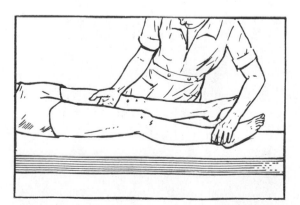

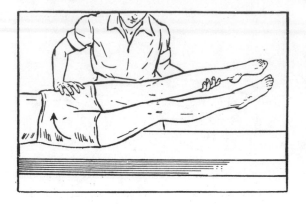

HIP ADDUCTION

Patient may attempt to substitute hip flexors for adductors by medially rotating hip and tipping pelvis backward toward the table or substitute hamstrings by laterally rotating hip and tipping pelvis forward. Sidelying position must be maintained.

PRIME MOVERS *(Continued)*

MUSCLE	ORIGIN	INSERTION
Adductor longus N: Obturator (anterior division L3, 4)	a. Anterior surface of pubis at angle of junction of crest with symphysis	a. Middle portion of medial lip of linea aspera
Pectineus N: Femoral (L2, 3, 4 and Accessory Obturator)	a. Pectineal line and area just anterior to it, between iliopectineal eminence and tubercle of pubis	a. Line between lesser trochanter and linea aspera
Gracilis N: Obturator (anterior division L3, 4)	a. Inferior half of symphysis pubis b. Superior half of pubic arch	a. Medial surface of tibia distal to condyle

HIP LATERAL ROTATION

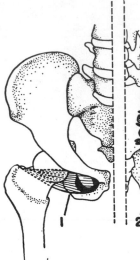
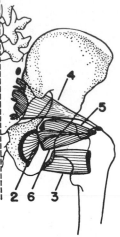

Anterior View
1. *Obturator externus*
2. *Obturator internus*
3. *Quadratus femoris*

Posterior View
4. *Piriformis*
5. *Gemellus superior*
6. *Gemellus inferior*
7. *Gluteus maximus*
(not illustrated. See Page 45.)

Range of Motion:

0 to 40–50 degrees (less with hip extended)

Factors Limiting Motion:

1. Tension of lateral band of iliofemoral ligament
2. Tension of hip medial rotator muscles

Measurement:

Sitting on side of table, hips and knees flexed about 90 degrees. Laterally rotate thigh. (See "Fair" test position.)

1. Place stationary arm of the goniometer perpendicular to the floor with the axis (fulcrum) at center of knee joint.
2. Movable arm is placed on anterior longitudinal midline of leg.

Patient grasps side of table to stabilize pelvis. Do not allow hip flexion, abduction, adduction or a shift of the trunk laterally.

PRIME MOVERS

MUSCLE	ORIGIN	INSERTION
Obturator externus N: Obturator (posterior division L3, 4)	a. Medial side of bony margin of obturator foramen b. Medial two-thirds of outer surface of obturator membrane c. Rami of pubis d. Ramus of ischium	a. Posterior surface of neck of femur to trochanteric fossa of femur
Obturator internus N: Nerve to the Obturator internus (L5, S1, 2)	a. Internal surface superior and inferior rami of the pubis b. Ramus of the ischium c. Pelvic surface of superior part of greater sciatic foramen and inferior and anterior obturator foramen	a. Through lesser sciatic notch to anterior part of medial surface of greater trochanter proximal to trochanteric fossa
Quadratus femoris N: Nerve to the Quadratus femoris (L4, 5, S1)	a. Proximal portion of external border of ischial tuberosity	a. Proximal part of linea quadrata of femur
Piriformis (S1, 2)	a. Anterior surface of sacrum between first and fourth anterior sacral foramina b. Margin of greater sciatic foramen and anterior surface of sacrotuberous ligament	a. Through greater sciatic foramen to superior border of greater trochanter of femur

(Continued on Page 60)

HIP LATERAL ROTATION

NORMAL AND GOOD

Sitting with legs over edge of table, pad under knee of limb to be tested.

Use counterpressure above knee to prevent abduction and flexion of hip. Patient grasps edge of table to stabilize pelvis.

Patient laterally rotates hip. Do not allow patient to assist lateral rotation by lifting pelvis on contralateral side of body or by attempting to flex knee or abduct hip.

Resistance is given above ankle joint.

Note: Resistance should be given slowly and carefully in tests for rotation of the hip and shoulder. Use of the long lever arm can cause injury to joint structures if not controlled.

If sitting position is contraindicated, tests may be given with patient supine as in test for Poor grade below. Resistance is given above knee and at ankle joint.

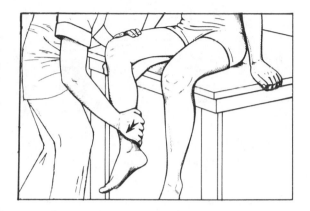

FAIR

Sitting with legs over edge of table, pad under knee of limb to be tested.

Use counterpressure above knee.

Patient laterally rotates hip through range of motion with stabilization of pelvis by patient.

POOR

Supine or standing with hip in medial rotation.

Stabilize pelvis.

Patient laterally rotates hip through range of motion.

In the supine position, slight resistance should be given during the last half of the range of motion as gravity assists in that part of the movement.

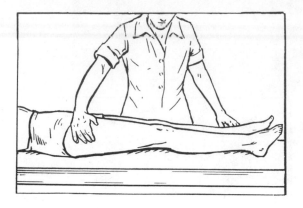

HIP LATERAL ROTATION

TRACE AND ZERO

Presence of contraction in lateral rotators may be determined by deep palpation primarily between the greater trochanter and the ischium as patient attempts lateral rotation of hip.

PRIME MOVERS *(Continued)*

MUSCLE	ORIGIN	INSERTION
Gemellus superior N: Branch of the nerve to the Obturator internus (L5, S1, 2)	a. Outer surface of ischial spine	a. Upper margin of tendon of Obturator internus and with it to medial surface of greater trochanter
Gemellus inferior (Branch of the nerve to the Quadratus femoris (L4, 5, S1)	a. Superior part of ischial tuberosity	a. Lower margin of tendon of Obturator internus and with it to medial surface of greater trochanter
Gluteus maximus (illustrated on page 44) N: Inferior gluteal (L5, S1, 2)	a. From posterior gluteal line and lateral lip of iliac crest superior and dorsal to line b. Posterior surface of lower part of sacrum and side of coccyx c. Posterior surface of sacrotuberous ligament and aponeurosis of the Erector spinae	a. Iliotibial band of fascia lata over greater trochanter b. Gluteal tuberosity

Accessory Muscles

Sartorius   Biceps femoris (long head)

FOR NOTES:

HIP MEDIAL ROTATION

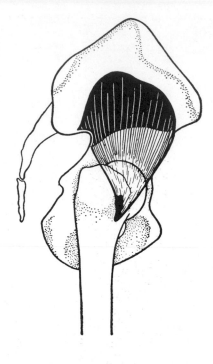

Lateral View
Gluteus minimus

Range of Motion:

0 to 35–45 degrees (less with hip extended)

Factors Limiting Motion:

1. When hip is extended, tension of iliofemoral ligament
2. When hip is flexed, tension of ischiocapsular ligament
3. Tension of hip lateral rotator muscles

Measurement:

Sitting on side of table, hips and knees flexed about 90 degrees. Medially rotate thigh. (See "Fair" test position.)

1. Place stationary arm of the goniometer perpendicular to the floor with the axis (fulcrum) at center of knee joint.
2. Movable arm is placed on anterior longitudinal midline of leg.

See Lateral Rotation for stabilization.

PRIME MOVERS

Muscle	Origin	Insertion
Gluteus minimus N: Superior gluteal (L4, 5, S1)	a. Outer surface of ilium between anterior and inferior gluteal lines b. Margin of greater sciatic notch	a. Anterior aspect of greater trochanter of femur b. Expansion to capsule of hip joint
Tensor fascia lata (illustrated on page 52) N: Superior gluteal (L4, 5, S1)	a. Anterior part of outer lip of iliac crest b. Outer surface of anterior superior iliac spine c. Deep surface of the fasciae latae	a. Between 2 layers of iliotibial band at juncture of middle and proximal thirds (iliotibial band inserts on lateral condyle of tibia)

Accessory Muscles
Gluteus medius (anterior fibers)
Semitendinosus
Semimembranosus

HIP MEDIAL ROTATION

NORMAL AND GOOD

Sitting with legs over edge of table, pad under knee of limb to be tested.

Use counterpressure above knee to prevent adduction of hip. (Patient grasps edge of table to stabilize pelvis.)

Patient medially rotates hip.

Resistance is given above ankle joint.

Note: In all tests, do not allow patient to assist medial rotation by lifting pelvis on side of limb being tested or attempt to extend knee or adduct and extend hip.

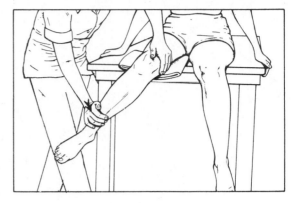

FAIR

Sitting with legs over edge of table, pad under knee of limb to be tested.

Use counterpressure above knee.

Patient medially rotates hip through range of motion with stabilization of pelvis.

POOR

Supine or standing with hip in lateral rotation.

Stabilize pelvis.

Patient medially rotates hip through range of motion.

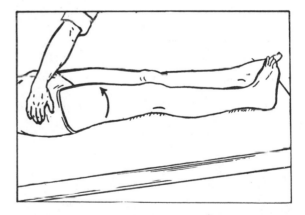

TRACE AND ZERO

Tensor fascia lata may be palpated near its origin posterior and distal to anterior superior spine of ilium. Gluteus minimus fibers lie beneath Gluteus medius and Tensor.

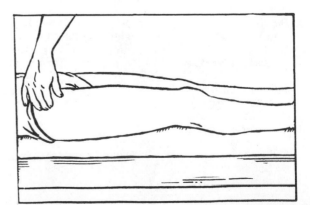

KNEE FLEXION

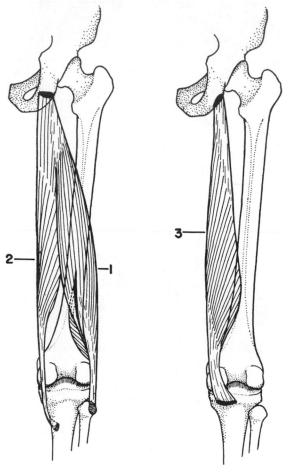

Range of Motion

0 to 135–145 degrees

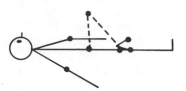

Factors Limiting Motion:

1. Tension of Quadriceps
2. Contact of calf with posterior thigh

Measurement:

Supine, hip flexed 90 degrees, knee completely flexed.

1. Place stationary arm of the goniometer on lateral longitudinal midline of thigh.
2. Movable arm is placed on lateral longitudinal midline of leg.

Do not allow hip abduction, adduction or rotation.

Posterior View
1. *Biceps femoris*
2. *Semitendinosus*
3. *Semimembranosus*

PRIME MOVERS

Muscle	Origin	Insertion
Biceps femoris (long head) N: Sciatic (tibial portion S1, 2, 3)	a. Distal and medial impression on ischial tuberosity b. Inferior part of sacrotuberous ligament	a. Lateral side of head of fibula b. Lateral condyle of tibia
Biceps femoris (short head) N: Sciatic (peroneal portion L5, S1, 2)	a. Lateral lip of linea aspera and prolongation to lateral condyle of femur	
Semitendinosus N: Sciatic (tibial portion L5, S1, 2)	a. Distal and medial impression on ischial tuberosity	a. Proximal part of anteromedial surface of tibia

(Continued on Page 66)

KNEE FLEXION

NORMAL AND GOOD

(Biceps femoris)

Prone with limbs straight.

Stabilize pelvis.

Patient flexes knee. Grasping above ankle, laterally rotate leg (muscle is placed in better alignment), and resist flexion to test Biceps femoris.

Note: Knee should be flexed less than 90 degrees if muscles tend to cramp.

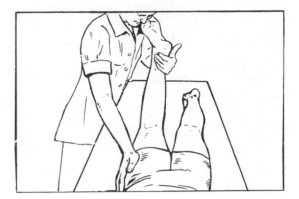

NORMAL AND GOOD

(Semitendinosus and Semimembranosus)

Prone with limbs straight.

Stabilize pelvis.

Patient flexes knee. Grasping proximal to the ankle, medially rotate leg and resist flexion to test Semimembranosus and Semitendinosus.

See Note above.

FAIR

Prone with limbs straight.

Stabilize thigh medially and laterally without pressure over the muscle group being tested.

Patient flexes knee to 90 degrees (remainder of motion is assisted by gravity).

Note: If Gastrocnemius is weak, knee may be flexed to 10 degrees for starting position.

If Biceps femoris is stronger, leg will laterally rotate during flexion.

If Semitendinosus and Semimembranosus are stronger, leg will medially rotate during flexion.

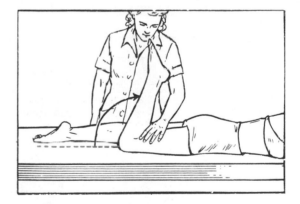

POOR

Sidelying with limbs straight and upper limb supported.

Stabilize thigh.

Patient flexes knee through range of motion.

Uneven muscular pull will cause rotation of leg as above.

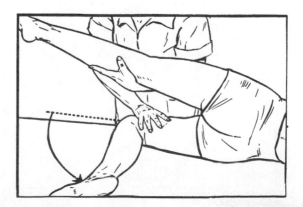

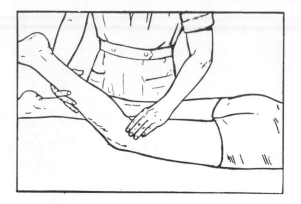

KNEE FLEXION

TRACE AND ZERO

Prone with knee partially flexed and leg supported.

Patient attempts to flex knee. Tendons of knee flexor muscles may be palpated on back of thigh near knee joint.

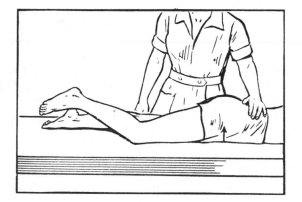

Patient may flex hip in order to start movement with knee partially flexed.

Note: The Sartorius may be substituted, which causes flexion and lateral rotation of the hip. Knee flexion in this position is less difficult, since leg is not raised vertically against gravity.

The Gracilis may be substituted, which causes hip adduction.

Strong plantar flexion of the foot should not be allowed in order to prevent substitution by the Gastrocnemius.

PRIME MOVERS (Continued)

MUSCLE	ORIGIN	INSERTION
Semimembranosus N: Sciatic (tibial portion L5, S1, 2)	a. Proximal and outer impression of ischial tuberosity	a. Horizontal groove on postero-medial aspect of medial condyle of tibia b. Tendon of insertion gives off fibrous expansion into posterior aspect of lateral femoral condyle

Accessory Muscles

Popliteus	Gracilis
Sartorius	Gastrocnemius

KNEE EXTENSION

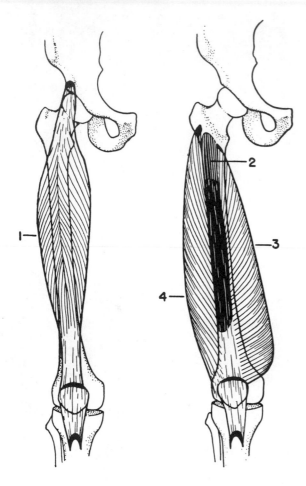

Range of Motion:

135–145 to 0 degrees

Factors Limiting Motion:

1. Tension of oblique popliteal, cruciate and collateral ligaments of knee joint
2. Tension of knee flexor muscles

Measurement:

Supine, knee and hip straight.

1. Place stationary arm of the goniometer on lateral longitudinal midline of thigh.
2. Movable arm is placed on lateral longitudinal midline of leg.

The heel may be raised from the table in the zero position because of the bulk of the gastrocsoleus. True hyperextension should be noted.

Anterior Views
Quadriceps femoris
1. *Rectus femoris*
2. *Vastus intermedius*
3. *Vastus medialis*
4. *Vastus lateralis*

PRIME MOVERS

MUSCLE	ORIGIN	INSERTION
Quadriceps femoris N: Femoral (L2, 3, 4)		
Rectus femoris	a. Anterior inferior spine of ilium (straight head) b. Groove just above brim of acetabulum (reflected head)	a. Base of patella b. Tendon of Quadriceps femoris into tuberosity of the tibia
Vastus intermedius	a. Anterior and lateral surfaces of proximal two-thirds of femoral shaft	a. Forms deep part of Quadriceps femoris tendon, inserting into base of patella b. Tendon of Quadriceps femoris into tuberosity of the tibia

(Continued on Page 70)

KNEE EXTENSION

NORMAL AND GOOD

Sitting with legs over edge of table. Patient grasps edge of table to stabilize trunk.

Patient should be allowed to lean back until tension in the hamstrings is relieved. Pain or discomfort in this muscle group will inhibit knee extension.

Stabilize thigh without pressure over Quadriceps.

Patient extends knee through range of motion without terminal locking.

Resistance is given above ankle joint. (Pad should be used under knee.)

Note: Resistance to a locked knee can be injurious to the joint and is not a valid indicator of the strength of the extensors because a co-contraction of other muscles around the knee is required for the locking action.

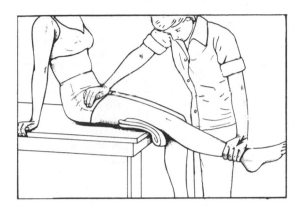

FAIR

Sitting with legs over edge of table.
Stabilize thigh.

Patient extends knee through range of motion without medial or lateral rotation at the hip (rotation allows extension at an angle, not in a vertical line against gravity).

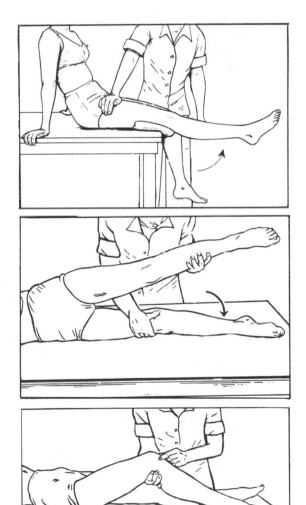

POOR

Sidelying with upper limb supported. Limb to be tested is flexed at the knee joint.

Stabilize thigh above knee joint. (Avoid pressure over Quadriceps.)

Patient extends knee through range of motion.

Note: Do not allow the hip to be in flexion as extending the hip from the flexed position can cause passive knee extension.

TRACE AND ZERO

Supine with knee flexed and supported.
Patient attempts to extend knee.

Contraction of Quadriceps is determined by palpation of tendon between patella and tuberosity of tibia, fibers of muscle on the anterior surface of the thigh, and the tendon of Rectus femoris near its origin between the Sartorius and the Tensor fascia lata.

MUSCLE	ORIGIN	INSERTION
Vastus medialis	a. Distal half of intertrochanteric line b. Medial lip of linea aspera and proximal part of medial supra-condylar ridge	a. Medial border of patella and Quadriceps femoris tendon and expansion to capsule of knee joint b. Tendon of the Quadriceps femoris into tuberosity of the tibia
Vastus lateralis	a. Proximal portion of intertrochan-teric line b. Anterior and inferior borders of greater trochanter c. Lateral lip of gluteal tuberosity d. Proximal half of lateral lip of linea aspera	a. Lateral border of patella, forming part of Quadriceps femoris tendon b. Tendon of the Quadriceps femoris into tuberosity of the tibia

ANKLE PLANTAR FLEXION

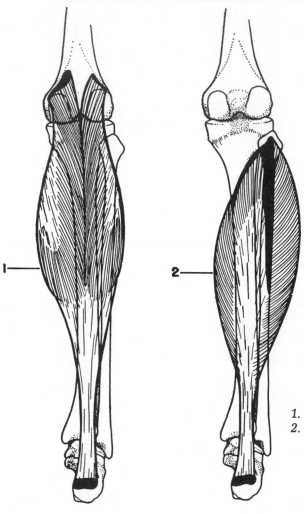

Posterior View

Range of Motion:

0 to 45–55 degrees

Factors Limiting Motion:

1. Tension on anterior talofibular ligament and anterior fibers of deltoid ligament
2. Tension of dorsiflexor muscles of ankle
3. Contact of posterior portion of talus with tibia

Measurement:

Supine, knee and hip straight.

1. Place stationary arm of the goniometer on lateral longitudinal midline of leg.
2. Movable arm is placed parallel to the fifth metatarsal with the axis (fulcrum) over lateral malleolus of fibula.

Do not allow inversion or eversion of the foot.

1. *Gastrocnemius*
2. *Soleus*

PRIME MOVERS

MUSCLE	ORIGIN	INSERTION
Gastrocnemius N: Tibial (S1, 2)	*Medial head:* a. Depression on proximal and posterior parts of medial condyle of femur and adjacent area *Lateral head:* a. Impression on side of lateral condyle and posterior surface of femur just proximal to it	a. Tendo calcaneus, which inserts into middle part of proximal surface of calcaneus
Soleus N: Tibial (S1, 2)	a. Posterior surface of head of fibula b. Proximal third of posterior surface of body of fibula c. Popliteal line and middle third of medial border of tibia	a. Tendo calcaneus

Accessory Muscles

Tibialis posterior	Peroneus brevis	Flexor digitorum longus
Peroneus longus	Flexor hallucis longus	Plantaris

ANKLE PLANTAR FLEXION

NORMAL AND GOOD

(for Gastrocnemius and Soleus)

Standing on limb to be tested, knee extended.

For Normal grade, patient raises heel from floor through range of motion of plantar flexion. Patient should complete motion four or five times in good form and without apparent fatigue.

The Tibialis posterior and the Peroneus longus and brevis muscles must be normal or good to stabilize the forefoot and provide counter pressure against the floor.

A Good grade is given if patient can complete full range of motion two or three times and then has difficulty in completing the movement.

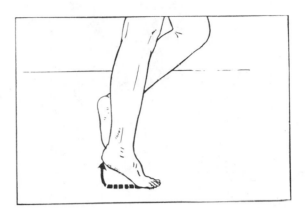

FAIR

Standing on limb to be tested. Knee extended.

Patient plantar flexes ankle sufficiently to clear heel from floor. (Not illustrated.)

POOR

Sidelying with limb to be tested resting on lateral surface, knee extended and ankle in midposition.

Stabilize leg.

Patient plantar flexes ankle through range of motion.

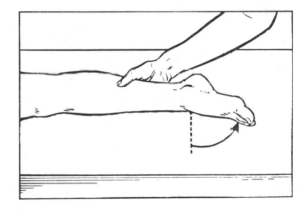

TRACE AND ZERO

Contraction of Gastrocnemius and Soleus is determined by palpation of tendon above calcaneus and muscle fibers on posterior aspect of leg. (Not illustrated.)

NORMAL AND GOOD

(for Soleus)

Standing on limb to be tested with knee slightly flexed.

For Normal grade, patient raises heel from floor through range of motion of plantar flexion maintaining flexed position of knee. Patient can complete motion four or five times in good form and without apparent fatigue.

A Good grade is given if patient can complete full range of motion two or three times and then has difficulty in completion.

Note: Flexion of the knee during the test allows "slack" in the Gastrocnemius, which originates proximal to the knee joint.

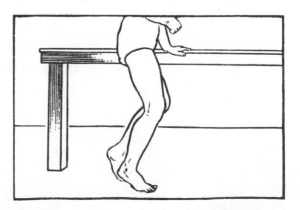

NONWEIGHT-BEARING TESTS

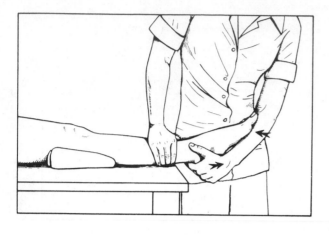

Supine with pad under knee to prevent hyperextension.

Stabilize leg proximal to ankle.

Patient plantar flexes ankle.

Resistance is given by grasping around the calcaneus and exerting pressure against the pull of the plantar flexors. Counterpressure may be given with the forearm against the sole of the foot if the accessory muscles that stabilize the forefoot are functioning.

Grades of Normal, Good and Fair may be based on amount of resistance given; however, they should be recorded with a question mark, since only weight-bearing tests give accurate grading of strength of Gastrocnemius and Soleus.

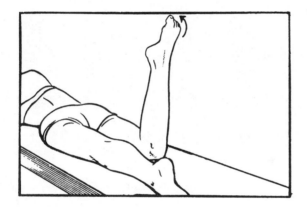

Substitution by the Flexor digitorum longus and Flexor hallucis longus will result in strong toe flexion with plantar flexion of forefoot and incomplete movement of calcaneus.

Substitution by Peroneus longus and Peroneus brevis will cause eversion of foot; substitution by Tibialis posterior will cause inversion of foot. Substitution by all three muscles will bring about plantar flexion of forefoot with limited movement of calcaneus.

FOOT DORSIFLEXION AND INVERSION

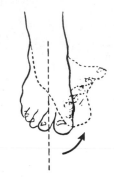

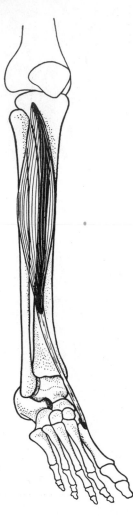

Anterolateral View
Tibialis anterior

Range of Motion:

0 to 15–25 degrees

Factors Limiting Motion:

1. Tension of lateral tarsal ligaments
2. Tension of Peroneus longus and Peroneus brevis muscles
3. Contact of tarsal bones medially

Measurement of Dorsiflexion Without Inversion:

Supine, knee partially flexed and supported by a pillow to lessen tension in the gastrocnemius.

1. Place stationary arm of the goniometer on lateral longitudinal midline of leg.
2. Movable arm is placed parallel to the fifth metatarsal with the axis (fulcrum) over lateral maleolus of fibula.

Do not allow inversion or eversion of foot.

PRIME MOVER

MUSCLE	ORIGIN	INSERTION
Tibialis anterior N: Deep peroneal (L4, 5, S1)	a. Lateral condyle and proximal two-thirds of anterolateral surface of tibial body b. Interosseous membrane	a. Medial and plantar surfaces of first cuneiform bone b. Base of first metatarsal bone

FOOT DORSIFLEXION AND INVERSION

NORMAL AND GOOD

Sitting with legs over edge of table.

Stabilize leg above ankle joint.

Patient dorsiflexes and inverts foot, keeping toes relaxed.

Resistance is given on medial dorsal aspect of foot.

Note: Patient should be cautioned to keep toes relaxed to avoid substitution by Extensors digitorum and Hallucis longus.

If the sitting position is contraindicated, tests may be given in the supine position. For a Fair grade, slight resistance may be applied.

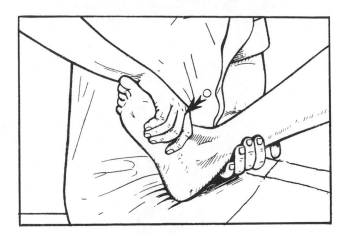

FAIR AND POOR

Sitting with legs over edge of table.

Stabilize leg above ankle joint.

Patient inverts and dorsiflexes foot through full range of motion for Fair grade or through partial range for Poor.

Note: Supine position may be used with a complete range of motion for a Poor grade.

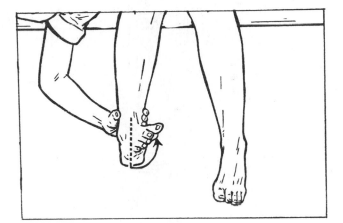

TRACE AND ZERO

Tendon of Tibialis anterior may be palpated on medial volar aspect of ankle and muscle fibers on the anterolateral portion of the leg (latter not illustrated).

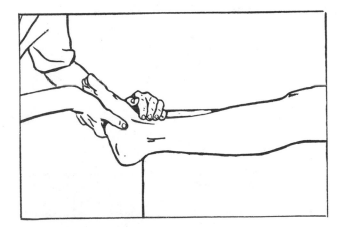

FOOT INVERSION

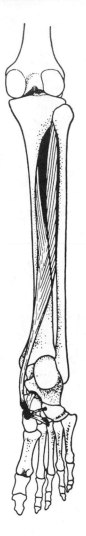

Posterior View of Leg
Plantar View of Foot
Tibialis posterior

Range of Motion:

0 to 30–40 degrees

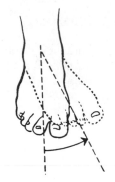

Factors Limiting Motion:

1. Tension of lateral tarsal ligaments
2. Tension of peroneal muscle group
3. Contact of tarsal bones medially

Measurement:

Sitting on side of table with knees flexed. Foot is inverted.

1. Place stationary arm of goniometer on anterior longitudinal midline of leg.
2. Movable arm is placed on the dorsum of the foot parallel to lateral side of second metatarsal.

The goniometer must be held forward and away from the ankle as it does not fit the contour of the joint.

PRIME MOVER

Muscle	Origin	Insertion
Tibialis posterior N: Tibial (L5, S1)	a. Proximal two thirds of medial surface of fibula b. Lateral portion of posterior surface of tibial body between beginning of popliteal line proximally and junction of middle and lower thirds of body distally c. Interosseous membrane	a. Tuberosity of navicular bone b. Fibrous expansions to sustentaculum tali of calcaneus, to 3 cuneiforms, cuboid, and bases of second, third, and fourth metatarsal bones

Accessory Muscles
Flexor digitorum longus
Flexor hallucis longus
Gastrocnemius (medial head)

FOOT INVERSION

NORMAL AND GOOD

Sidelying, ankle in slight plantar flexion.

Stabilize leg. Avoid pressure over Tibialis posterior muscle.

Patient inverts foot through range of motion.

Resistance is given on medial border of forefoot.

Flexors of the toes should remain relaxed to prevent substitution by the Flexor digitorum longus and Flexor hallucis longus.

Note: Inversion combines supination, adduction, and plantar flexion.

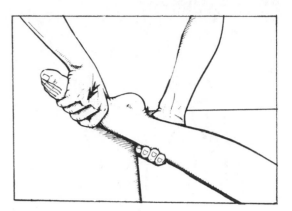

FAIR

Sidelying, ankle in slight plantar flexion.

Stabilize leg. Avoid pressure over Tibialis posterior muscle.

Patient inverts foot through range of motion.

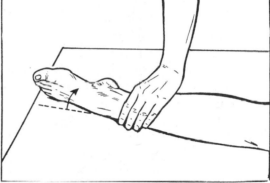

POOR

Supine, ankle in slight plantar flexion.
Stabilize leg.
Patient inverts foot through range of motion.

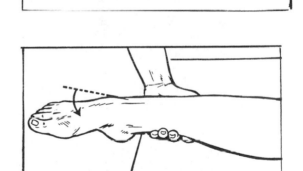

TRACE AND ZERO

Tendon of Tibialis posterior may be found between medial malleolus and navicular bone. It is also palpable above malleolus.

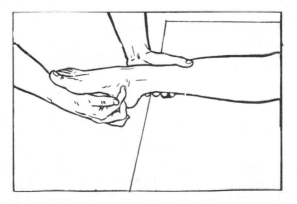

FOOT EVERSION

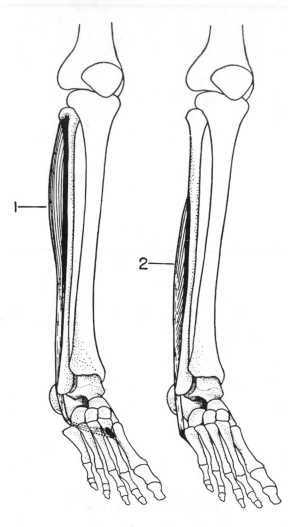

Anterolateral View
1. *Peroneus longus*
2. *Peroneus brevis*

Range of Motion:

0 to 15–25 degrees

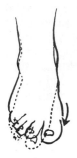

Factors Limiting Motion:

1. Tension of medial tarsal ligaments
2. Tension of Tibialis anterior and Tibialis posterior muscles
3. Contact of tarsal bones laterally

Measurement:

Sitting on side of table with knees flexed. Foot is everted.

1. Place stationary arm of goniometer on anterior longitudinal midline of leg.
2. Movable arm is placed on the dorsum of the foot parallel to lateral side of second metatarsal.

The goniometer must be held forward and away from the ankle as it does not fit the contour of the joint.

PRIME MOVERS

MUSCLE	ORIGIN	INSERTION
Peroneus longus N: Superficial peroneal (L4, 5, S1)	a. Head and proximal two thirds of lateral surface of body of fibula b. Occasionally a few fibers from lateral condyle of tibia	a. Tendon passes behind lateral malleolus, obliquely forward to groove on inferior surface of cuboid and sole of foot to lateral side of base of first metatarsal and medial cuneiform bones
Peroneus brevis N: Superficial peroneal (L4, 5, S1)	a. Distal two thirds of lateral surface of body of fibula	a. Behind lateral malleolus to tuberosity of base of fifth metatarsal on lateral side of bone

Accessory Muscles

Extensor digitorum longus   Peroneus tertius

Foot Eversion

NORMAL AND GOOD

Sidelying, ankle in midposition between plantar and dorsiflexion.

Stabilize leg.

Patient everts foot and depresses head of first metatarsal.

To test Peroneus brevis, resistance is given on lateral border of foot.

To test Peroneus longus, resistance is given against plantar surface of first metatarsal head.

The two muscles may be tested together using a derotating motion as illustrated.

Note: Eversion is a combination of pronation, abduction and dorsiflexion.

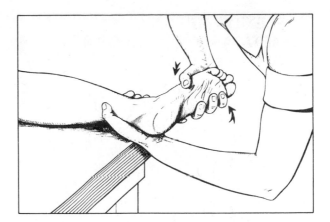

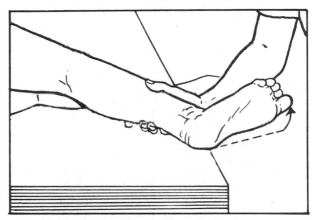

FAIR

Sidelying, ankle in midposition between plantar and dorsiflexion.

Stabilize leg.

Patient everts foot through range of motion and depresses first metatarsal.

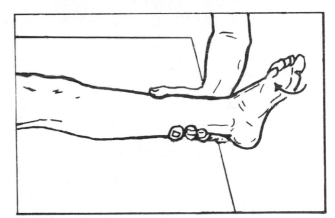

POOR

Supine, ankle in midposition between plantar and dorsiflexion.

Stabilize leg.

Patient everts foot through range of motion with depression of first metatarsal.

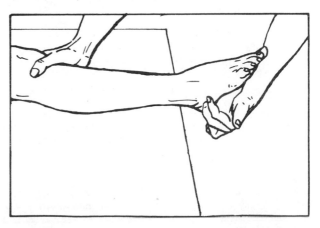

TRACE AND ZERO

Tendon of Peroneus brevis may be found proximal to base of fifth metatarsal on lateral border of foot and bulk of muscle on lower portion of lateral surface of leg over fibula.

Contraction in Peroneus longus may be determined by palpation over the upper half of the lateral surface of the leg distal to the head of the fibula. (Not illustrated.)

FLEXION OF METATARSOPHALANGEAL JOINTS OF TOES

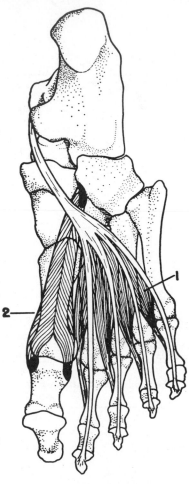

Range of Motion:

0 to 35–45 degrees

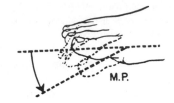

Factors Limiting Motion:

1. Tension of extensor muscle tendons of toes
2. Contact of soft parts

Measurement:

Supine, ankle dorsiflexed to lessen tension in the long extensors of toes. Metatarsophalangeal joints of the lateral four toes and hallux are flexed.

1. Place stationary arm of goniometer on the longitudinal midline of the dorsal surface of each metatarsal of the joint to be measured.
2. Movable arm is placed on the midline of dorsal surface of proximal phalanx.

Use a small goniometer.

Plantar View
1. *Lumbricals*
2. *Flexor hallucis brevis*

Flexion of Metatarsophalangeal Joints of Lateral Four Toes

PRIME MOVER

Muscle	Origin	Insertion
Lumbricals First lumbrical: N: Medial plantar (L4, 5) Second, third and fourth lumbricals: N: Lateral plantar (S1, 2)	a. Tendons of Flexor digitorum longus, as far back as their divisions	a. Pass distally around medial sides of lateral 4 toes to insert into expansions of tendons of Extensor digitorum longus on dorsal surfaces of proximal phalanges

Accessory Muscles
Dorsal and plantar interossei
Flexor digiti quinti brevis
Flexor digitorum longus
Flexor digitorum brevis

(Continued on Page 83)

FLEXION OF METATARSOPHALANGEAL JOINTS OF TOES

FLEXION OF METATARSOPHALANGEAL JOINTS OF LATERAL 4 TOES (LUMBRICALS)

Supine, ankle in midposition between plantar and dorsiflexion.

Stabilize metatarsals.

Patient flexes lateral 4 toes at the metatarsophalangeal joints, keeping interphalangeal joints extended.

Resistance is given on the plantar surface of the proximal row of phalanges for grades of Normal and Good.

Extension of the middle and distal interphalangeal joints with the metacarpophalangeal joints in flexion should be tested concurrently. Extension of these joints is part of the primary action of the Lumbricals as in the hand.

Resistance also should be given to each toe separately as Lumbricals are uneven in strength and have divided innervation.

Range of motion is completed for a Fair grade, a partial range for Poor.

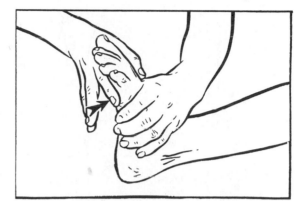

FLEXION OF METATARSOPHALANGEAL JOINT OF HALLUX (FLEXOR HALLUCIS BREVIS)

Supine, ankle in midposition between plantar and dorsiflexion.

Stabilize first metatarsal.

Patient flexes hallux. Resistance is given beneath proximal phalanx for Normal and Good.

Note: Grades below the level of normal and good may be difficult to determine because joint motion is often limited and the muscle and tendon cannot be palpated. If range appears normal, a grade of Fair may be given for completion of full motion, a Poor for partial range.

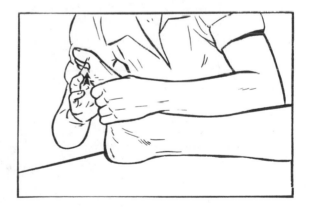

Flexion of Metatarsophalangeal Joint of Hallux

PRIME MOVER

MUSCLE	ORIGIN	INSERTION
Flexor hallucis brevis N: Medial plantar (L4, 5, S1)	a. Medial part of plantar surface of cuboid b. Contiguous portion of lateral cuneiform bone	a. By 2 tendons into medial and lateral sides of base of proximal phalanx of hallux (sesamoid bone in each tendon)

Accessory Muscle
Flexor hallucis longus

FLEXION OF INTERPHALANGEAL JOINTS OF TOES

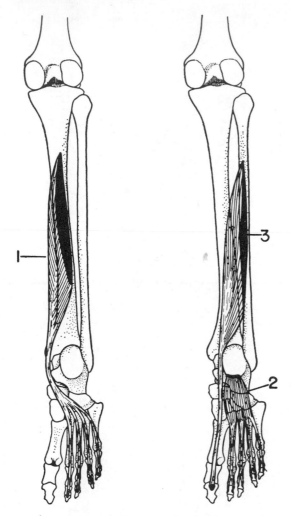

Posterior View of Leg
Plantar View of Foot

Range of Motion:

0 to 50–80 degrees

Factors Limiting Motion:

1. Tension of extensor muscle tendons (dorsal ligaments) of toes
2. Contact of soft parts of phalanges

Measurement:

Supine, ankle dorsiflexed to lessen tension in long extensors of toes. Interphalangeal joints of lateral four toes are flexed.

The midline on the dorsal surface of the proximal, middle, and distal phalanges are used for placement of the goniometer.

The interphalangeal joint of the hallux is measured in the same way as the smaller toes.

Use a small goniometer.

1. *Flexor digitorum longus*
2. *Flexor digitorum brevis*
3. *Flexor hallucis longus*

Flexion of Proximal Interphalangeal Joints of Lateral 4 Toes

PRIME MOVER

MUSCLE	ORIGIN	INSERTION
Flexor digitorum brevis N: Medial plantar (L4, 5)	a. Medial process of tuberosity of calcaneus	a. By 4 tendons that divide into two slips and insert into sides of second phalanges of lateral 4 toes

Flexion of Distal Interphalangeal Joints of Lateral 4 Toes

PRIME MOVER

MUSCLE	ORIGIN	INSERTION
Flexor digitorum longus N: Tibial (L5, S1)	a. Posterior surface of body of tibia distal to the popliteal line to within 7 or 8 cm of its distal end	a. Bases of distal phalanges of lateral 4 toes

(Continued on Page 85)

FLEXION OF INTERPHALANGEAL JOINTS OF TOES

FLEXION OF PROXIMAL INTERPHALANGEAL JOINTS OF LATERAL 4 TOES (FLEXOR DIGITORUM BREVIS)

Supine, ankle in midposition between plantar and dorsiflexion.

Stabilize proximal row of phalanges of lateral 4 toes.

Patient flexes toes.

Resistance is given beneath middle row of phalanges of lateral 4 toes for Normal and Good.

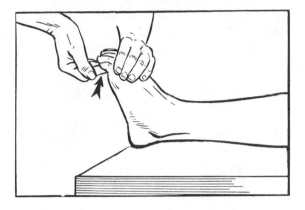

FLEXION OF DISTAL INTERPHALANGEAL JOINTS OF LATERAL 4 TOES (FLEXOR DIGITORUM LONGUS)

Supine, ankle in midposition between plantar and dorsiflexion.

Stabilize middle row of phalanges of lateral 4 toes.

Patient flexes toes.

Resistance is given beneath distal row of phalanges of lateral 4 toes for Normal and Good.

A grade of Fair may be given for completion of the range of motion and Poor for partial range in all tests of toe flexion.

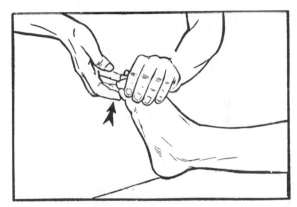

FLEXION OF INTERPHALANGEAL JOINT OF HALLUX (FLEXOR HALLUCIS LONGUS)

Supine, ankle in midposition between plantar and dorsiflexion.

Stabilize proximal phalanx of hallux.

Patient flexes toe.

Resistance is given beneath distal phalanx for Normal and Good.

For Trace and Zero, tendon of Flexor hallucis longus may be found on plantar surface of proximal phalanx.

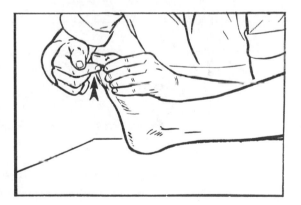

Flexion of Interphalangeal Joint of Hallux

PRIME MOVER

MUSCLE	ORIGIN	INSERTION
Flexor hallucis longus N: Tibial (L5, S1, 2)	a. Inferior two thirds of posterior surface of body of fibula b. Lower part of interosseous membrane	a. Base of terminal phalanx of great toe

EXTENSION OF METATARSOPHALANGEAL JOINTS OF TOES AND INTERPHALANGEAL JOINT OF HALLUX

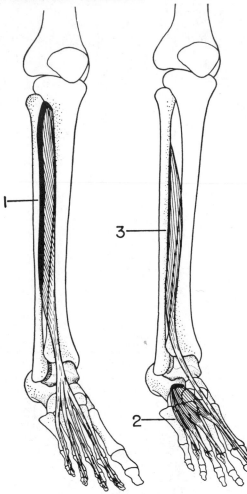

Anterolateral View

Range of Motion:

0 to 75–85 degrees

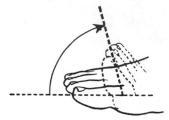

Factors Limiting Motion:

Tension of plantar and collateral ligaments of toe joints.

Measurement:

Supine, ankle plantar flexed to lessen tension in the long flexors of the toes. Metatarsophalangeal joints of the lateral four toes and hallux are extended.

1.   Place stationary arm of the goniometer on plantar surface of foot in line with each metatarsal.
2.   Movable arm is placed on the midline of plantar surface of proximal phalanx.

1.   *Extensor digitorum longus*
2.   *Extensor digitorum brevis*
3.   *Extensor hallucis longus*

Extension of Metatarsophalangeal Joints of Toes and Interphalangeal Joint of Hallux
PRIME MOVERS

MUSCLE	ORIGIN	INSERTION
Extensor digitorum longus N: Deep peroneal (L4, 5, S1)	a. Lateral condyle of tibia b. Proximal three-fourths of anterior surface of fibula	a. Extensor expansions into middle and distal phalanges of lateral 4 toes
Extensor digitorum brevis N: Deep peroneal (L5, S1)	a. Distal and lateral surfaces of calcaneus in front of groove for Peroneus brevis muscle	a. Medial division into dorsal surface of proximal phalanx of hallux at base (sometimes called Extensor hallucis brevis) b. Three lateral divisions into lateral sides of tendon of Extensor digitorum longus of second, third, and fourth toes

(Continued on Page 88)

EXTENSION OF METATARSOPHALANGEAL JOINTS OF TOES AND INTERPHALANGEAL JOINT OF HALLUX

EXTENSION OF METATARSOPHALANGEAL JOINTS OF LATERAL 4 TOES (EXTENSOR DIGITORUM LONGUS AND EXTENSOR DIGITORUM BREVIS)

Supine, ankle in midposition between plantar and dorsiflexion.

Stabilize metatarsal area.

Patient extends lateral 4 toes.

Resistance is given on proximal phalanges of toes for Normal and Good.

A grade of Fair may be given for completion of the range of motion and Poor for partial range in all tests for toe extension.

For Trace and Zero grades, tendons of Extensor digitorum longus may be palpated on dorsal surface of metatarsals and fibers of Extensor digitorum brevis on lateral side of dorsum of foot anterior to malleolus.

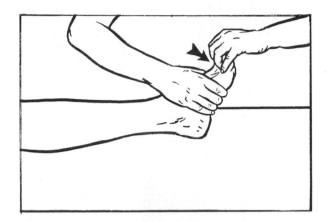

EXTENSION OF METATARSOPHALANGEAL JOINT OF HALLUX (MEDIAL DIVISION OF EXTENSOR DIGITORUM BREVIS)

Supine, ankle in midposition between plantar and dorsiflexion.

Stabilize first metatarsal.

Patient extends metatarsophalangeal joint of hallux.

Resistance is given over proximal phalanx for Normal and Good.

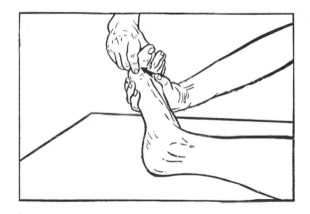

EXTENSION OF INTERPHALANGEAL JOINT OF HALLUX (EXTENSOR HALLUCIS LONGUS)

Supine, ankle in midposition between plantar and dorsiflexion.

Stabilize proximal phalanx of hallux.

Patient extends distal joint of hallux.

Resistance is given on dorsal surface for Normal and Good.

For Trace and Zero grades, tendon of Extensor hallucis longus may be palpated over the dorsal surface of first metatarsophalangeal joint and on a diagonal line across dorsum of foot to middle of anterior aspect of ankle.

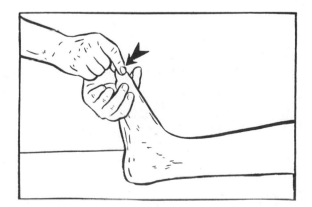

Extension of Interphalangeal Joint of Hallux

PRIME MOVER

MUSCLE	ORIGIN	INSERTION
Extensor hallucis longus N: Deep peroneal (L4, 5, S1)	a. Middle two-fourths of anterior surface of fibula	a. Base of distal phalanx of hallux

INNERVATION OF THE MUSCLES OF THE UPPER LIMB
(Showing Derivation from the 11th Cranial Nerve, Cervical and the Brachial Plexuses)

CRANIAL NERVE AND VENTRAL PRIMARY DIVISIONS OF SPINAL NERVES			Connecting Limb to Vertebral Column and Thoracic Wall			
Roots	Trunks	Cords	Branches	Arm	Forearm	Hand

Hand

MEDIAN (C6. 7)
Abductor pollicis brevis
Opponens pollicis
Flexor pollicis brevis (lateral)
1st and 2nd lumbricals

DEEP PALMAR BRANCH (C8. T1)
Flexor pollicis brevis (medial)
Adductor pollicis
  Oblique head
  Transverse head
3rd and 4th lumbricals (C8)
Dorsal and Palmar interossei

ULNAR (C8. T1)
Abductor digiti minimi
Flexor digiti minimi brevis
Opponens digiti minimi
Palmaris brevis (C8)

(Nerves are in CAPITAL LETTERS)

Forearm

MEDIAN
Pronator teres (C6. 7)
Flexor carpi radialis (C6. 7)
Palmaris longus (C6. 7)
Flexor digitorum superficialis (C7. 8. T1)

PALMAR INTEROSSEOUS (C8. T1)
Flexor digitorum profundus
Flexor pollicis longus
Pronator quadratus

RADIAL
Brachioradialis (C5. 6)
Extensor carpi radialis longus (C6. 7)
Extensor carpi radialis brevis (C6. 7)
Ancoreus (C7. 8)

DEEP RADIAL (C6. 7. 8)
Extensor digitorum
Extensor digiti minimi
Extensor carpi ulnaris
Supinator (C6)
Adductor pollicis longus (C6. 7)
Extensor pollicis longus
Extensor pollicis brevis (C6. 7)
Extensor indicis

ULNAR (C8. T1)
Flexor carpi ulnaris
Flexor digitorum profundus

Arm

Coracobrachialis (C6. 7)
Biceps brachii (C5. 6)
Brachialis (C5. 6)

Triceps brachii (C7. 8)

MUSCULOCUTANEUS (C5. 6. 7)

MEDIAN

RADIAL (C5. 6. 7. 8. T1)

ULNAR (C8. T1)

(C6. 7. 8. T1)

Branches / Cords / Trunks

LATERAL

POSTERIOR

MEDIAL

SUPERIOR

MIDDLE

INFERIOR

(a) through (m) labels

Roots

Cr 11 ACCESSORY
C3
C4
C5
C6
C7
C8
T1

(a) SPINAL ACCESSORY Cr. 11)
  Trapezius
(b) BRANCH TO LEVATOR SCAPULAE (C3)
(c) BRANCH TO LEVATOR SCAPULAE (C4)
(c) DORSAL SCAPULAR (C5)
  Rhomboid major
  Rhomboid minor
  Levator scapulae (frequent)
(e) LONG THORACIC (C5–7)
  Serratus anterior

(f) NERVE TO SUBCLAVIUS (C5. 6)
(g) SUPRASCAPULAR (C5. 6)
  Supraspinatus
  Infraspinatus
(h) SUPERIOR SUBSCAPULAR (C5. 6)
  Subscapularis
(i) INFERIOR SUBSCAPULAR (C5. 6)
  Subscapularis
  Teres major

(j) THORACODORSAL (C6–8)
  Latissimus dorsi
  Pectoralis major
(k) LATERAL PECTORAL (C5–7)
(l) AXILLARY (C5. 6)
  Deltoid
  Teres minor (C5)
(m) MEDIAL PECTORAL (C8. T1)
  Pectoralis major
  Pectoralis minor

—Chart adapted from Worthingham C *Upper and Lower Extremity Muscle and Innervation Charts.* Stanford University Press. 1944.

89

SCAPULAR ABDUCTION AND UPWARD ROTATION

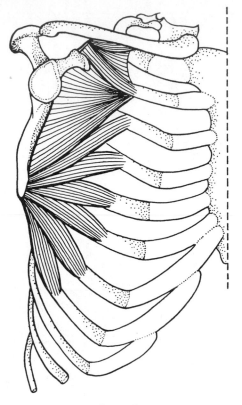

Anterolateral View
Serratus anterior

*Range of Motion:**

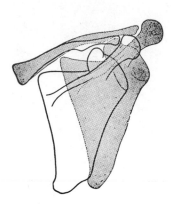

Factors Limiting Motion:

　　1.　Tension of trapezoid ligament (limits forward rotation of scapula upon clavicle)
　　2.　Tension of Trapezius and Rhomboideus major and minor muscles.

Fixation:

　　1.　In strong scapular abduction, pull of Obliquus externus abdominis on same side
　　2.　Weight of thorax

---

　*Drawn from x-ray views of subject at beginning and end of movement in test for normal muscle. Change in shape of scapula is due to shift in plane during movement.

PRIME MOVER

MUSCLE	ORIGIN	INSERTION
Serratus anterior N: Long thoracic （C5, 6, 7)	a. Digitations from outer surfaces and superior borders of upper 8 or 9 ribs b. Aponeuroses covering inter- costal muscles	a. Ventral surface of superior angle of scapula b. Ventral surface of vertebral border of scapula c. Lower 5 or 6 digitations con- verge and insert on ventral surface of inferior angle of scapula

NORMAL AND GOOD

Supine, shoulder flexed to 90 degrees with slight abduction, and elbow in extension.

Patient moves arm upward by abducting the scapula.

Resistance is given by grasping around forearm and elbow. Pressure is downward and inward toward table.

Observe scapula for "winging" (movement of vertebral border away from thorax and substitution by anterior muscles of shoulder).

FAIR

Supine, shoulder flexed to 90 degrees and scapula resting on table.

Stabilize thorax.

Patient forces arm upward. Scapula should be completely abducted without "winging."

If extensor muscles of elbow are weak, elbow may be flexed, or forearm may be supported.

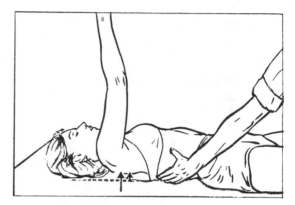

POOR

Sitting with shoulder flexed to 90 degrees and arm resting on table.

Stabilize thorax.

Patient moves arm forward by abducting scapula.

Note: If the sitting position is contraindicated, tests may be given in the supine position. Patient completes partial range of motion for a Poor grade.

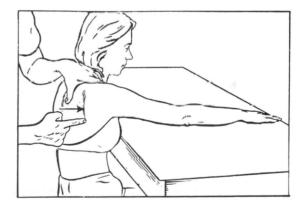

TRACE AND ZERO

Examiner lightly forces arm backward to determine presence of a contraction of Serratus anterior. Scapula should be observed for "winging."

Digitations of the Serratus anterior may be palpated along midaxillary line and lower between External oblique and Latissimus dorsi. (Latter not illustrated.)

SCAPULAR ELEVATION

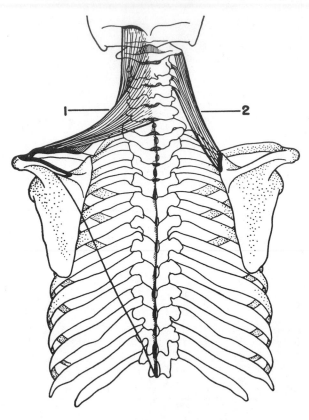

Posterior View
1. *Trapezius (superior fibers)*
2. *Levator scapulae*

Range of Motion:*

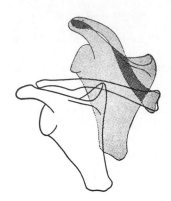

Factors Limiting Motion:

1. Tension of costoclavicular ligament
2. Tension of muscles depressing scapula and clavicle: Pectoralis minor, Subclavius, and Trapezius (lower fibers)

Fixation:

1. Flexor muscles of cervical spine (for tests done in sitting position)
2. Weight of head (for tests done in prone position)

*Drawn from x-ray views of subject at beginning and end of movement in test for normal muscle.

PRIME MOVERS

MUSCLE	ORIGIN	INSERTION
Trapezius (superior fibers) N: Accessory (spinal portion Cr. 11)	a. External occipital protuberance b. Medial third of superior nuchal line of occipital bone c. Upper part of ligamentum nuchae	a. Posterior border of lateral third of clavicle
Levator scapulae N: (C3, 4) and frequently branch from dorsal scapular (C5)	a. Transverse processes of atlas, axis, and third and fourth cervical vertebrae	a. Vertebral border of scapula superior to root of spine
Accessory Muscles Rhomboid major and minor		

SCAPULAR ELEVATION

NORMAL AND GOOD

Sitting with arms relaxed at sides.

Patient raises shoulders as high as possible.

Resistance is given downward on top of shoulders.

Note: If the sitting position is contraindicated, tests for Normal, Good, and Fair may be given in the supine position. Slight resistance is given for a Fair grade.

FAIR

Sitting with arms relaxed at sides.
Patient elevates shoulders through range of motion.

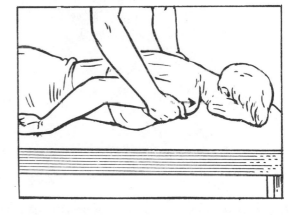

POOR

Prone with shoulders supported by examiner and forehead resting on table. (Prone position is preferable because examiner can see any unevenness in muscle contraction.)

Patient moves shoulders toward ears through range of motion.

Note: If the prone position is uncomfortable, tests for Poor, Trace, and Zero may be given in the supine position.

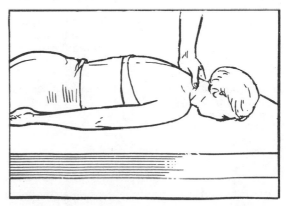

TRACE AND ZERO

Prone.
Examiner palpates upper fibers of Trapezius parallel to cervical vertebrae and near insertion above clavicle as patient attempts to elevate shoulders.

SCAPULAR ADDUCTION

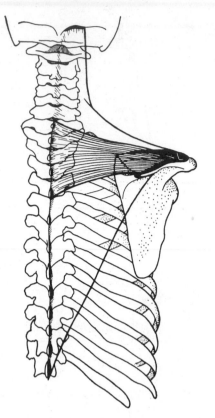

Posterior View
Trapezius (middle fibers)

*Range of Motion:**

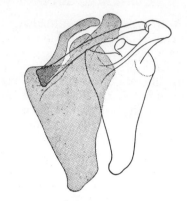

Factors Limiting Motion:

1. Tension of conoid ligament (limits backward rotation of scapula upon clavicle)
2. Tension of Pectoralis major and minor and Serratus anterior muscles
3. Contact of vertebral border of scapula with spinal musculature

Fixation:

Weight of trunk

---

*Drawn from x-ray studies of subject at beginning and end of movement in test for normal muscles.

PRIME MOVERS

MUSCLE	ORIGIN	INSERTION
Trapezius (middle fibers) N: Accessory (spinal portion Cr. 11)	a. Inferior part of ligamentum nuchae b. Spinous processes of seventh cervical and superior thoracic vertebrae (horizontal fibers of muscle)	a. Medial margin of acromion process of scapula b. Superior lip of posterior border of spine of scapula
Rhomboid major and *minor* (illustrated on page 100) N: Dorsal scapular (C5)	a. Spinous processes of seventh cervical and first 5 thoracic vertebrae	a. Vertebral border of scapula between root of spine and inferior angle

Accessory Muscles
Trapezius (lower and upper fibers)

SCAPULAR ADDUCTION

NORMAL AND GOOD

Prone, shoulder abducted to 90 degrees and laterally rotated, elbow flexed to a right angle.

Stabilize thorax.

Patient horizontally abducts arm and adducts scapula.

Resistance is given on lateral angle of scapula. (No pressure is placed on the humerus.)

Note: If glenohumeral muscles are weak, arm may be placed in a vertical position over edge of table.

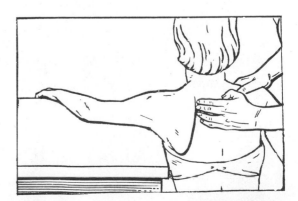

FAIR

Prone, shoulder abducted to 90 degrees and laterally rotated, elbow flexed to a right angle.

Stabilize thorax.

Patient raises arm and adducts scapula.

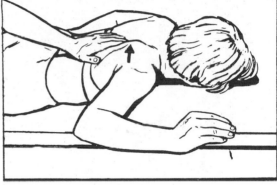

POOR

Sitting, arm supported in 90 degrees of abduction, elbow partially flexed.

Stabilize thorax.

Patient horizontally abducts arm and adducts scapula.

Note: If the sitting position is contraindicated, tests for Poor, Trace, and Zero may be given in the prone position. Patient completes partial range of motion for a Poor grade.

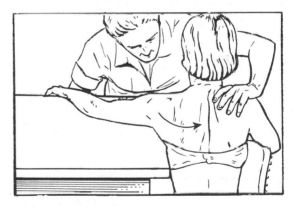

TRACE AND ZERO

Sitting or prone.

Middle fibers of Trapezius are palpated above spine of scapula from acromion to vertebral column to determine presence of a contraction.

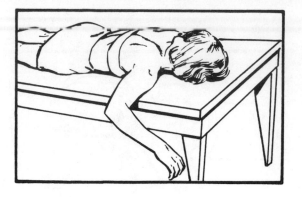

SCAPULAR ADDUCTION

Depression of shoulder toward table with scapular abduction on attempt to perform test motion indicates that posterior fibers of Deltoid are contracting, but scapula is not fixed or adducted.

SCAPULAR DEPRESSION AND ADDUCTION

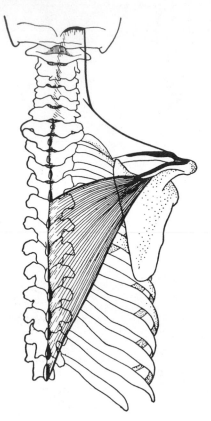

Posterior View
Trapezius (inferior fibers)

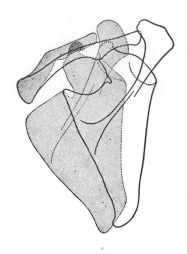

*Range of Motion:**

Factors Limiting Motion:

   1.   Tension of interclavicular ligament and articular disk of sternoclavicular joint
   2.   Tension of Trapezius (upper fibers), Levator scapulae, and Sternocleidomastoid (clavicular head)

Fixation:

   1.   Contraction of spinal extensor muscles
   2.   Weight of trunk

---

*Drawn from x-ray views of subject at beginning and end of movement in test for normal muscle.

PRIME MOVER

Muscle	Origin	Insertion
Trapezius (inferior fibers) N: Accessory (spinal portion Cr. 11)	a. Spinous processes of inferior thoracic vertebrae and corresponding supraspinal ligament	a. By aponeurosis gliding over medial end of spine of scapula to tubercle at apex of smooth triangular surface

Accessory Muscles
Trapezius: (middle fibers, adduction)

NORMAL AND GOOD

Prone, patient's head rotated to opposite side and shoulder to be tested in approximately 130 degrees of abduction.

Patient raises arm and fixes scapula strongly with lower part of Trapezius.

Resistance is given on lateral angle of scapula in upward and outward direction.

Note: If range of motion in shoulder is limited, arm may be placed over side of table and supported at maximal range of examiner.

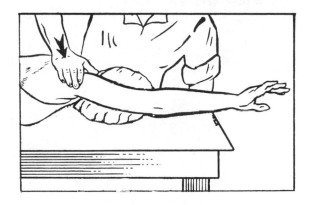

NORMAL AND GOOD

(Alternate)

If Deltoid is weak, arm is passively raised by examiner.

Patient attempts to assist. Resistance is given on scapula as above.

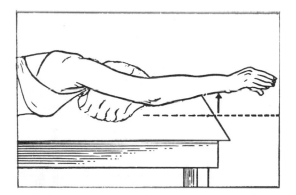

FAIR AND POOR

Prone, patient's head rotated to opposite side and shoulder to be tested in approximately 130 degrees of abduction.

Patient lifts arm from table through full range of motion without upward movement of the scapula or forward sagging of the acromion process for Fair grade or through partial range for Poor.

If Deltoid is weak, grade is based on amount and firmness of contraction of lower fibers of Trapezius.

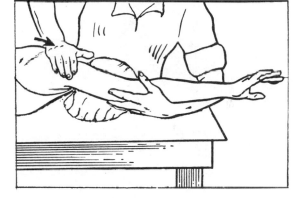

TRACE AND ZERO

Lower fibers of Trapezius are palpated in triangular area between the root of the spine of the scapula and the lower half of the thoracic vertebrae as patient attempts to lift arm from table.

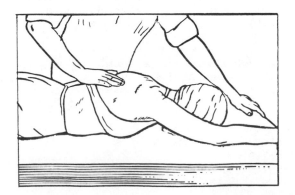

SCAPULAR ADDUCTION AND DOWNWARD ROTATION

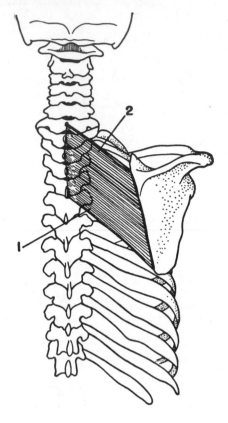

Range of Motion:*

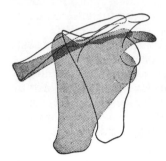

Factors Limiting Motion:

1. Tension of conoid ligament (limits backward rotation of scapula upon clavicle)
2. Tension of Pectoralis major and minor and Serratus anterior muscles
3. Contact of vertebral border of scapula with spinal musculature

Fixation:

Weight of trunk

Posterior View
1. *Rhomboid major*
2. *Rhomboid minor*

---

*Drawn from x-ray views of subject at beginning and end of movement in test for normal muscles.

PRIME MOVERS

MUSCLE	ORIGIN	INSERTION
Rhomboid major N: Dorsal scapular (C5)	a. Spinous processes of second, third, fourth, and fifth thoracic vertebrae	a. Tendinous arch from root of spine of scapula to inferior angle (arch connected to scapula by thin membrane)
Rhomboid minor N: Dorsal scapular (C5)	a. Inferior part of ligamentum nuchae b. Spinous processes of seventh cervical and first thoracic vertebrae	a. Base of triangular smooth surface at root of spine of scapula
Accessory Muscle Trapezius (adduction)		

SCAPULAR ADDUCTION AND DOWNWARD ROTATION

NORMAL AND GOOD

Prone, patient's head rotated to opposite side.

Shoulder medially rotated and arm adducted across back. Shoulders relaxed.

Stabilize thorax.

Patient raises arm and adducts scapula.

Resistance is given on vertebral border of scapula in outward and slightly upward direction.

Note: Do not allow patient to force head of humerus downward against table in an attempt to lift arm. Both arm and scapula should move together.

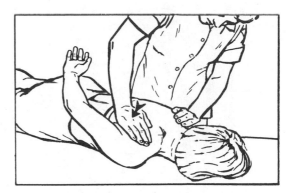

FAIR

Prone, patient's head rotated to opposite side.

Stabilize thorax.

Patient raises arm and adducts scapula.

If the glenohumeral muscles are weak, slight resistance may be given to the scapula for a Fair grade.

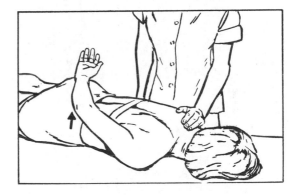

POOR

Sitting with shoulder medially rotated and arm adducted behind back. Stabilize trunk with anterior and posterior pressure to prevent flexion and rotation.

Patient adducts scapula.

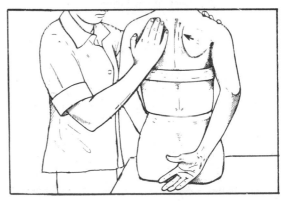

TRACE AND ZERO

Examiner palpates Rhomboid muscles at the angle formed by the vertebral border of the scapula and the lateral fibers of the lower Trapezius as patient attempts to adduct scapula.

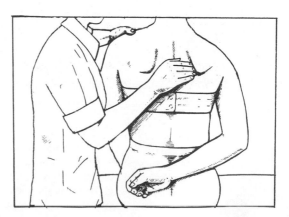

SHOULDER FLEXION

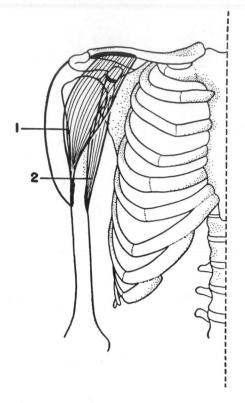

Anterior View
1. Deltoid (anterior fibers)
2. Coracobrachialis

Range of Motion:

0 to 170–180 degrees

Factors Limiting Motion:

1.   Contact of greater tubercle of humerus with anterior surface of acromion
2.   Shoulder extensor muscles

Measurement:

Supine, knees and hips flexed to lessen extension of the lumbar spine, elbow in full extension to minimize tension in the long head of the triceps. Shoulder flexed.

1.   Place stationary arm of goniometer on the midaxillary line of the thorax.
2.   Movable arm is placed on the lateral longitudinal midline of the arm.

Scapula must move anteriorly and laterally to allow full range of motion.

Do not allow the lower ribs to be lifted from table. With stabilization of the scapula, the above procedure may be used to isolate and measure glenohumeral motion.

PRIME MOVERS

MUSCLE	ORIGIN	INSERTION
Deltoid (anterior fibers) N: Axillary (C5, 6)	a. Anterior border and superior surface of lateral third of clavicle	a. Deltoid prominence on middle of the lateral side of body of humerus
Coracobrachialis N: Musculocutaneous (C6, 7)	a. Apex of coracoid process	a. Medial surface and border of humerus opposite insertion of Deltoid

Accessory Muscles
Deltoid (middle fibers)
Pectoralis major (clavicular fibers)
Biceps brachii
Serratus anterior
Trapezius

NORMAL AND GOOD

Sitting, arm at side with elbow slightly flexed to prevent lateral rotation at the shoulder joint and substitution by the Biceps.

Patient flexes shoulder to 90 degrees without rotation or horizontal movement.

If scapular muscles are weak, stabilize scapula; if normal, stabilize thorax.

Resistance is given proximal to the elbow joint.

Note: If the scapula is completely fixed by stabilization at the beginning of the movement, the strength of the Deltoid will be diminished. The scapula should be allowed to abduct and upwardly rotate according to the normal ratio of movement: After the first 20 degrees it is 2–1; this is 2 degrees in the glenohumeral joint to 1 degree in the scapulothoracic articulation.

FAIR AND POOR

Sitting position with arm at side and elbow slightly flexed.

Patient flexes shoulder to 90 degrees (palm down) for Fair and through partial range for Poor.

See stabilization above.

Alternate test for Poor grade: Sidelying, upper arm tested on a powdered board. Arm is flexed through a 90 degree range of motion.

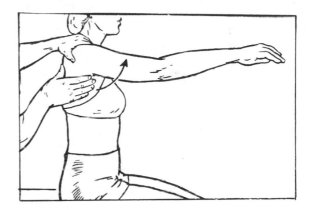

TRACE AND ZERO

Supine.

Fibers of the anterior section of the Deltoid are palpated on the anterior surface of the upper third of the arm as patient attempts to flex shoulder. The Coracobrachialis is deep lying in the upper medial third.

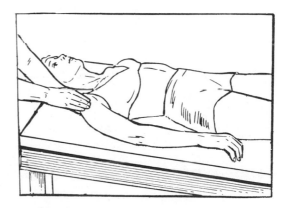

SHOULDER FLEXION

Patient may laterally rotate and attempt to flex shoulder with Biceps brachii. Arm should be kept in midposition between medial and lateral rotation.

Patient may lean back or try to elevate shoulder girdle to aid in flexion.

SHOULDER EXTENSION

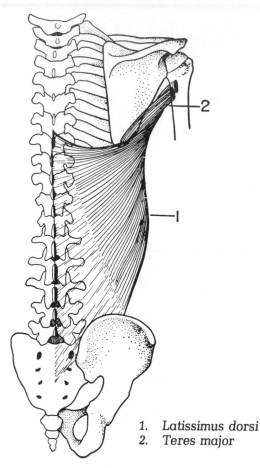

Posterior View

1. Latissimus dorsi
2. Teres major

Range of Motion:

0 to 50–60 degrees

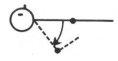

Factors Limiting Motion:

1. Tension of shoulder flexor muscles
2. Contact of greater tubercle of humerus with acromion posteriorly

Measurement:

Supine, arm over side of table, elbow flexed to minimize tension in the biceps muscle. Shoulder extended.

1. Place stationary arm of goniometer on the midaxillary line of the thorax.
2. Movable arm is placed on the lateral longitudinal midline of the arm.

Do not allow flexion of the thoracic spine or abduction at the shoulder joint.

PRIME MOVERS

Muscle	Origin	Insertion
Latissimus dorsi N: Thoracodorsal (C6, 7, 8)	a. Posterior layer of lumbodorsal fascia by which it is attached to spines of the lower 6 thoracic, lumbar and sacral vertebrae, supraspinal ligament, and to posterior iliac crest b. External lip of iliac crest lateral to Erector spinae c. Caudal 3 or 4 lower ribs d. Usually a few fibers from inferior angle of scapula	a. Bottom of intertubercular groove of humerus
Teres major N: Inferior subscapular (C5, 6)	a. Dorsal surface of inferior angle of scapula	a. Crest below lesser tuberosity of humerus posterior to Latissimus dorsi
Deltoid (posterior fibers illustrated on page 110) N: Axillary (C5, 6)	a. Inferior lip of posterior border of spine of scapula	a. Deltoid prominence on middle of the lateral side of body of humerus

Accessory Muscles

Teres minor   Triceps brachii (long head)

SHOULDER EXTENSION

NORMAL AND GOOD

Prone, shoulder medially rotated and adducted (palm up to prevent lateral rotation).

Patient extends shoulder through range of motion.

If scapular muscles are weak, stabilize scapula; if normal, stabilize thorax. Allow scapula to tip forward normally to complete range of motion.

Resistance is given proximal to the elbow joint. (Examiner should stand close to patient and on same side of table as extremity being tested.)

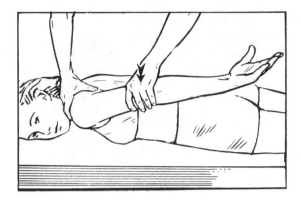

FAIR AND POOR

Prone, arm at side, shoulder medially rotated.

Patient extends shoulder through range of motion for Fair and partial range for Poor.

Alternate test for Poor grading: Sidelying, upper arm tested on a powdered board. Shoulder is extended through range of motion.

TRACE AND ZERO

Prone.

Examiner palpates the fibers of the Teres major on the lower border of the scapula, the fibers of the Latissimus dorsi slightly below, and the posterior fibers of the Deltoid along the posterior aspect of the arm.

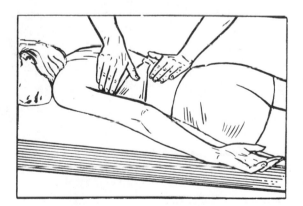

SHOULDER ABDUCTION

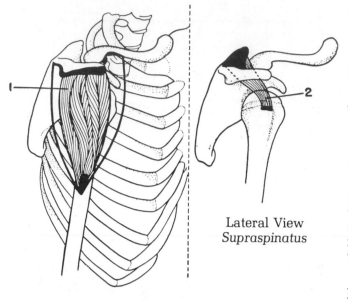

Lateral View
Supraspinatus

Lateral View
Deltoid (middle fibers)

Range of Motion:

0 to 170–180 degrees

Factors Limiting Motion:

1. Contact of greater tubercle of humerus with lateral surface of acromion
2. Shoulder adductor muscles

Measurement:

Supine shoulder abducted with lateral rotation of the humerus so that the greater tubercle may glide under rather than strike against the acromion. Elbow is extended to minimize tension in the long head of the triceps.

1. Place stationary arm of goniometer on lateral aspect of the thorax parallel to the sternum.
2. Movable arm is placed on the medial longitudinal midline of the arm.

Do not allow lateral trunk flexion to the opposite side.

PRIME MOVERS

MUSCLE	ORIGIN	INSERTION
Deltoid (middle fibers) N: Axillary (C5, 6)	a. Lateral margin and superior surface of acromion	a. Deltoid tubercle on middle of the lateral side of body of humerus
Supraspinatus N: Suprascapular (C5)	a. Medial two-thirds of supraspinatus fossa	a. Highest of 3 impressions on greater tubercle of humerus

Accessory Muscles
Deltoid (anterior and posterior fibers)
Serratus anterior (direct action on scapula)

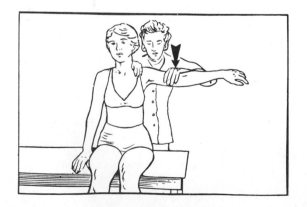

SHOULDER ABDUCTION

NORMAL AND GOOD

Sitting, arm at side and elbow flexed a few degrees.
Patient abducts shoulder 90 degrees.
Resistance is given proximal to elbow joint.
See Note about stabilization and ratio of movement under Shoulder Flexion, Page 103.

SHOULDER ABDUCTION

FAIR

Sitting, arm at side and elbow flexed a few degrees to prevent shoulder outward rotation and substitution by the Biceps.

Patient abducts shoulder 90 degrees.

See stabilization under Shoulder Flexion, Page 103.

Note: Do not allow elevation of the shoulder or lateral flexion of the trunk to the contralateral side. Both of these give the appearance of abduction.

POOR

Supine, arm at side with elbow slightly flexed.

Patient abducts shoulder 90 degrees without rotation.

Consider ratio of movement in stabilization.

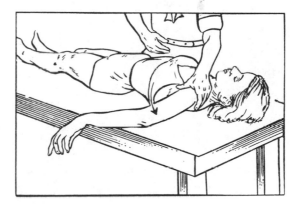

TRACE AND ZERO

Fibers of the middle section of the Deltoid may be palpated below the acromion process on the lateral surface of the upper third of the arm. The Supraspinatus lies under the Trapezius in the supraspinous fossa of the scapula.

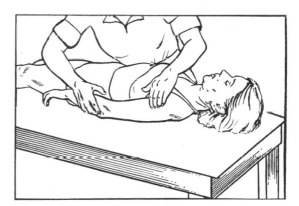

SHOULDER HORIZONTAL ABDUCTION

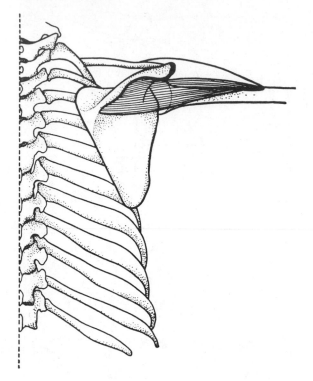

Posterior View
Deltoid (posterior fibers)

Range of Motion:

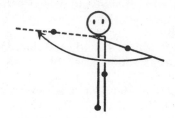

Factors Limiting Motion:

    1.   Tension of anterior fibers of capsule of glenohumeral joint
    2.   Tension of Pectoralis major and Deltoid (anterior fibers)

Fixation:

    Contraction of Rhomboid major and minor and Trapezius (primarily middle and lower fibers)

PRIME MOVER

Muscle	Origin	Insertion
Deltoid (posterior fibers) N: Axillary (C5, 6)	a. Inferior lip of posterior border of spine of scapula	a. Deltoid prominence on middle of the lateral side of body of humerus

Accessory Muscles
Infraspinatus
Teres minor

SHOULDER HORIZONTAL ABDUCTION

NORMAL AND GOOD

Prone, shoulder abducted to 90 degrees with forearm hanging vertically over edge of table.

Patient horizontally abducts shoulder through range of motion.

If scapular muscles are weak, stabilize scapula; if normal, stabilize thorax. (Examiner should stand close to patient and on same side of table as extremity being tested.)

Note: Do not allow extension of the elbow, which denotes substitution by the long head of the Triceps.

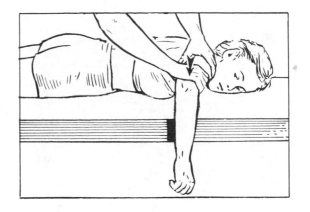

FAIR

Prone, shoulder abducted to 90 degrees with forearm hanging vertically over edge of table.

Patient horizontally abducts shoulder through range of motion.

See stabilization above.

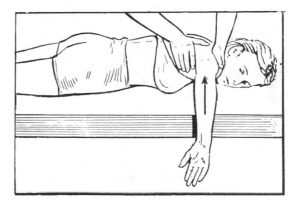

POOR

Sitting with arm supported in 90 degrees of flexion, elbow partially flexed.

Patient horizontally abducts shoulder through range of motion.

See stabilization above.

TRACE AND ZERO

Fibers of the posterior section of the Deltoid may be palpated below and lateral to the spine of the scapula and on the posterior aspect of the upper third of the arm.

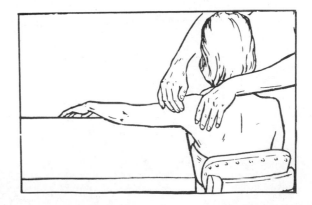

SHOULDER HORIZONTAL ADDUCTION

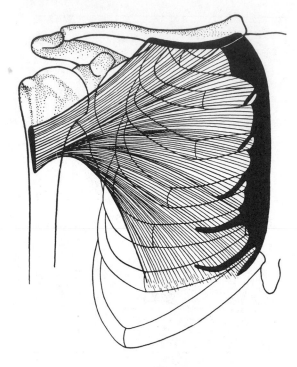

Anterior View
Pectoralis major

Range of Motion:

Factors Limiting Motion:

1. Tension of shoulder extensor muscles
2. Contact of arm with trunk.

Fixation:

In forceful horizontal adduction, contraction of External abdominal oblique muscle on same side

PRIME MOVER

MUSCLE	ORIGIN	INSERTION
Pectoralis major N: Medial and lateral pectoral (C5, 6, 7, 8, T1)	a. Anterior surface of sternal half of clavicle b. Half of breadth of ventral surface of sternum as far as sixth or seventh rib c. Cartilages of first 6 or 7 ribs	a. Crest of greater tubercle of humerus

Accessory Muscle
Deltoid (anterior fibers)

SHOULDER HORIZONTAL ADDUCTION

NORMAL AND GOOD

Supine, shoulder abducted to 90 degrees.

Stabilize thorax. (It may be necessary to stabilize thorax on contralateral side to prevent trunk from rolling to test side.)

Patient horizontally adducts shoulder through range of motion.

Resistance is given proximal to elbow joint.

Note: Sternal and clavicular portions of Pectoralis major may be isolated to some degree. In Normal and Good tests resistance is given in a direction opposite to line of pull of fibers (upward and outward for sternal part and downward and outward for clavicular part).

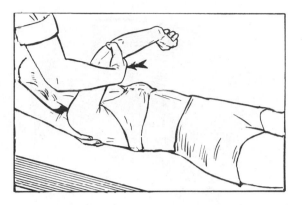

FAIR

Supine, shoulder abducted to 90 degrees. Stabilize thorax.

Patient adducts shoulder to vertical position.

Note: The arm may be placed above 90 degrees of abduction to test sternal portion of the Pectoralis major and below 90 degrees to test clavicular portion. Arm is then raised to the vertical in the direction of the muscle fibers being tested.

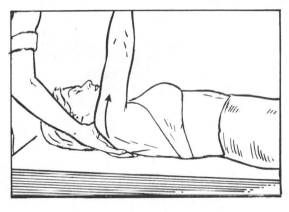

POOR

Sitting with arm supported on table with shoulder abducted to 90 degrees, elbow partially flexed.

Stabilize thorax.

Patient horizontally adducts shoulder through range of motion.

TRACE AND ZERO

Examiner palpates tendon of Pectoralis major near insertion on anterior aspect of upper arm. Muscle fibers of both sternal and clavicular portions may be observed and palpated on upper anterior aspect of thorax.

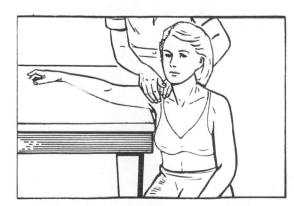

SHOULDER LATERAL ROTATION

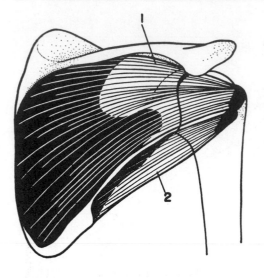

Dorsal View
1. *Infraspinatus*
2. *Teres minor*

Range of Motion:

0 to 80–90 degrees

Factors Limiting Motion:

1. Tension of superior portion of capsular ligament and coracohumeral ligament
2. Tension of medial rotator muscles of shoulder

Measurement:

Supine, shoulder abducted to 90 degrees, elbow flexed to 90, forearm in midposition. Shoulder laterally rotated.

1. Place stationary arm of goniometer perpendicular to the floor. Axis of motion is at the olecranon process.
2. Movable arm is placed on the lateral longitudinal midline of the forearm.

Thorax should not be raised from the table. Examiner should steady the patient with a hand on his elbow before the measurement is made.

PRIME MOVERS

Muscle	Origin	Insertion
Infraspinatus N: Suprascapular (C5, 6)	a. Medial two-thirds of infraspinatus fossa	a. Middle impression on greater tubercle of humerus
Teres minor N: Axillary (C5)	a. Cranial two-thirds of axillary border of scapula on dorsal surface	a. Most inferior of 3 impressions on greater tubercle of humerus and area just distal to it, uniting with posterior portion of capsule of shoulder joint

Accessory Muscle
Deltoid (posterior fibers)

NORMAL AND GOOD

Prone, shoulder abducted to 90 degrees, arm resting on table and forearm hanging vertically over edge. Small pillow or folded towel under arm.

Patient moves forearm upward through range of lateral rotation.

If scapular muscles are weak, stabilize scapula; if normal, stabilize thorax.

Resistance is given on forearm proximal to wrist.

Note: Resistance should be given slowly and carefully in tests for rotation at the hip and shoulder. Use of the long lever arm can cause injury to joint structures if not controlled.

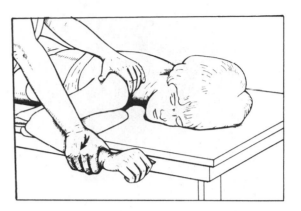

FAIR

Prone, shoulder abducted to 90 degrees, arm resting on table and forearm hanging vertically over edge.

Patient moves forearm upward through range of lateral rotation.

See stabilization above and place hand against anterior surface of arm to prevent shoulder abduction (without interfering with motion).

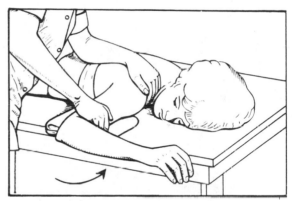

POOR

Prone with entire arm over edge of table in medially rotated position.

Patient laterally rotates shoulder through range of motion.

See stabilization above.

Note: Supination of the forearm should not be allowed to substitute for full range in lateral rotation.

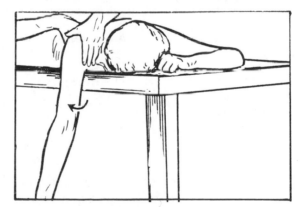

TRACE AND ZERO

The Teres minor may be palpated on axillary border of scapula, and Infraspinatus over body of scapula below the spine in the infraspinous fossa.

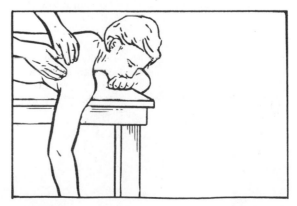

SHOULDER MEDIAL ROTATION

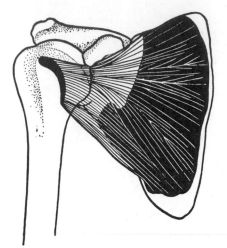

Costal View
Subscapularis

Range of Motion:

0 to 70–80 degrees

Factors Limiting Motion:

1. Tension of superior portion of capsular ligament
2. Tension of lateral rotator muscles of shoulder

Measurement:

Supine, shoulder abducted to 90 degrees, elbow flexed to 90, forearm in midposition. Shoulder medially rotated.

1. Place stationary arm of goniometer perpendicular to floor. Axis of motion is at the olecranon process.
2. Movable arm is placed on the lateral longitudinal midline of the forearm.

Thoracic spine should not be flexed. (See Lateral Rotation for additions.)

PRIME MOVERS

MUSCLE	ORIGIN	INSERTION
Subscapularis N: Superior and inferior subscapular (C5, 6)	a. Medial two-thirds of costal surface of scapula b. Inferior two-thirds of groove on axillary border of scapula	a. Lesser tubercle of humerus b. Ventral portion of capsule of shoulder joint
Pectoralis major (illustration on page 112) N: Medial and lateral pectoral (C5, 6, 7, 8, T1)	a. Anterior surface of sternal half of clavicle b. Half of breadth of ventral surface of sternum as far caudalward as seventh rib c. Cartilages of first 6 or 7 ribs	a. Crest of greater tubercle of humerus
Latissimus dorsi (illustration on page 106) N: Thoracodorsal (C6, 7, 8)	a. Spinous processes of lower 6 thoracic vertebrae b. Posterior layer of lumbodorsal fascia by which it is attached to spines of lumbar and sacral vertebrae, supraspinal ligament, and to posterior iliac crest c. External lip of iliac crest lateral to Erector spinae d. Three of 4 lower ribs e. Usually a few fibers from inferior angle of scapula	a. Bottom of intertubercular groove of humerus
Teres major (illustration on page 106) N: Inferior subscapular (C5, 6)	a. Dorsal surface of inferior angle of scapula	a. Crest below lesser tuberosity of humerus (posterior to Latissimus dorsi)

Accessory Muscle
Deltoid (anterior fibers)

SHOULDER MEDIAL ROTATION

NORMAL AND GOOD

Prone, shoulder abducted to 90 degrees, arm resting on table and forearm hanging vertically over edge. Small pillow or folded towel under arm.

Patient moves forearm upward through range of medial rotation.

If scapular muscles are weak, stabilize scapula; if normal, stabilize thorax. Allow scapula to tip forward normally to complete range of motion.

Resistance is given proximal to wrist.

See Note about resistance under Shoulder Lateral Rotation.

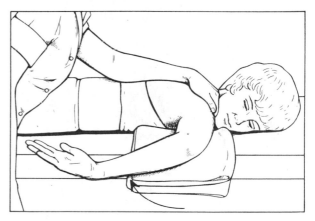

FAIR

Prone, shoulder abducted to 90 degrees, arm resting on table and forearm hanging vertically over edge. Small pillow or folded towel under arm.

Patient moves forearm upward through range of medial rotation.

See stabilization above, and if patient tends to adduct shoulder, place hand against posterior surface of arm (without interfering with motion).

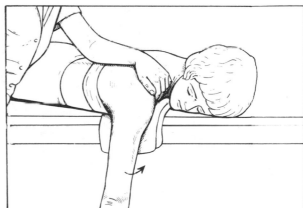

POOR

Prone with arm over edge of table in lateral rotation.

Patient medially rotates shoulder through range of motion.

See stabilization above.

Pronation of the forearm should not be allowed to substitute for full range in medial rotation.

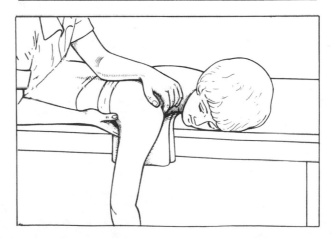

TRACE AND ZERO

Fibers of Subscapularis may be palpated deep in axilla near insertion.

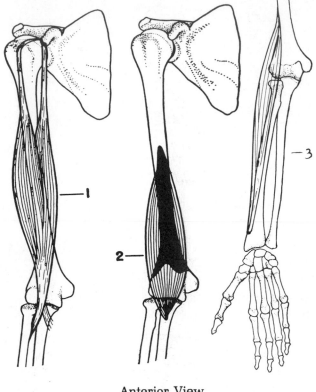

Anterior View
1. *Biceps brachii*
2. *Brachialis*
3. *Brachioradialis*

Range of Motion:

0 to 145–155 degrees

Factors Limiting Motion:

1. Contact of muscle masses on volar aspect of arm and forearm
2. Contact of coronoid process with coronoid fossa of humerus

Measurement:

Supine, arm at side and forearm in supination. Elbow flexed.

1. Place stationary arm of goniometer on the lateral longitudinal midline of the arm.
2. Movable arm is placed on the lateral longitudinal midline of the forearm.

PRIME MOVERS

Muscle	Origin	Insertion
Biceps brachii N: Musculocutaneous (C5, 6)	*Short head:* a. Flattened tendon from apex of coracoid process of scapula *Long head:* a. Tendon from supraglenoid tuberosity of scapula	a. Passes through capsule of shoulder joint and down in inter-tubercular groove to insert on posterior portion of tuberosity of radius and aponeurosis m. bicipitis brachii
Brachialis N: Musculocutaneous (C5, 6) and usually filament from radial	a. Distal half of anterior aspect of humerus	a. Tuberosity of ulna and anterior surface of coronoid process
Brachioradialis N: Radial (C5, 6)	a. Proximal two thirds of lateral supracondylar ridge of humerus b. Lateral intermuscular septum distal to groove for radial nerve	a. Flat tendon into lateral side of base of styloid process of radius

Accessory Muscles
Other flexor muscles (arising from medial epicondyle of humerus)

NORMAL AND GOOD

Sitting with arm at side and forearm supinated for test of Biceps, pronated for Brachialis, and in midposition for Brachioradialis.

Stabilize arm without pressure over Biceps or Brachialis.

Patient flexes elbow through range of motion.

Resistance is given proximal to wrist joint.

Note: The wrist flexors may be contracted for assistance in elbow flexion. Wrist will be strongly flexed as a result. Muscles of the wrist should be relaxed.

If sitting position is contraindicated, all tests may be given in the supine position. Slight resistance is given for a Fair grade.

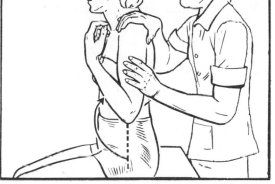

FAIR

Sitting with arm at side and forearm supinated.

Stabilize arm.

Patient flexes elbow through range of motion.

POOR

Supine, shoulder abducted to 90 degrees and laterally rotated.

Stabilize arm.

Patient slides forearm along table through complete range of elbow flexion.

Alternate test: Sitting, shoulder abducted 90 degrees with arm and forearm supported on table. Elbow extended and forearm supinated. Flex elbow through range of motion.

TRACE AND ZERO

Tendon of the Biceps may be palpated in the antecubital space; muscle fibers are found on the middle two-thirds of the anterior surface of the arm, short head medial to long head.

The fibers of the Brachialis may be palpated medial to lower section of Biceps and the Brachioradialis on the anterolateral surface of forearm below elbow. (Not illustrated.)

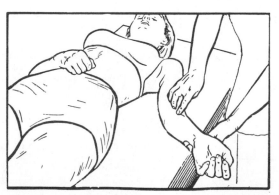

ELBOW EXTENSION

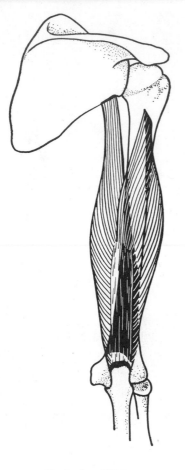

Posterior View
Triceps brachii

Range of Motion:

145–155 to 0 degrees

Factors Limiting Motion:

1. Tension of anterior, radial, and ulnar collateral ligaments of elbow joint
2. Tension of flexor muscles of forearm
3. Contact of olecranon process with olecranon fossa on posterior aspect of humerus

Measurement:

Supine, arm at side, and forearm in zero position between supination and pronation to partially eliminate the "carrying angle" at the elbow. Elbow extended.

1. Place stationary arm of the goniometer on the lateral longitudinal midline of the arm.
2. Movable arm is placed on the longitudinal midline of the dorsal surface of the forearm.

Record hyperextension if present.

PRIME MOVERS

Muscle	Origin	Insertion
Triceps brachii N: Radial (C7, 8)	*Long head:* a. Infraglenoid tuberosity of scapula *Lateral head:* a. Posterior surface of body of humerus proximal to groove for radial nerve *Medial head (Deep head):* a. Posterior surface of body of humerus distal to groove for radial nerve	a. Posterior portion of proximal surface of olecranon b. Fibrous expansion to deep fascia of forearm

Accessory Muscles
Anconeus
Extensor muscles of forearm (arising from lateral condyle of humerus)

ELBOW EXTENSION

NORMAL AND GOOD

Supine, shoulder flexed to 90 degrees and elbow flexed.

Stabilize arm.

Patient extends elbow through range of motion.

Resistance is given proximal to wrist joint in plane of forearm motion. Elbow should be in slight flexion and not in locked extension. (See Knee Extension.)

Do not allow arm to rotate, as resistance should be in line with axis of the joint.

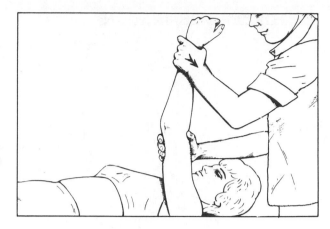

FAIR

Supine, shoulder flexed to 90 degrees and elbow flexed.

Stabilize arm.

Patient extends elbow through range of motion.

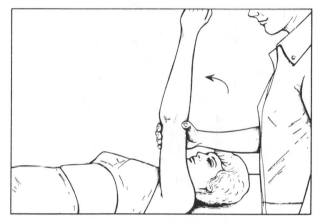

POOR

Supine, shoulder abducted to 90 degrees and laterally rotated. Elbow is flexed.

Stabilize arm.

Patient extends elbow through range of motion.

See Elbow Flexion for alternate test.

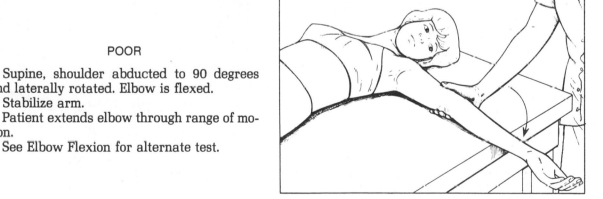

TRACE AND ZERO

Tendon of the Triceps may be palpated proximal to the olecranon process and muscle fibers on the posterior aspect of the arm.

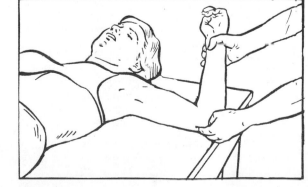

FOREARM SUPINATION

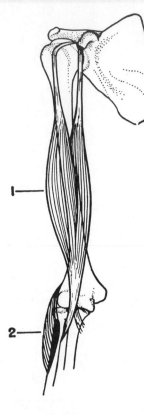

Anterior View
1. *Biceps brachii*
2. *Supinator*

Range of Motion:

0 to 80–90 degrees

Factors Limiting Motion:

1. Tension of volar radioulnar ligament and ulnar collateral ligament of wrist joint
2. Tension of oblique cord and lowest fibers of interosseous membrane
3. Tension of pronator muscles of forearm

Measurement:

Sitting, elbow flexed to 90 degrees, arm against thorax. Forearm pronated.

1. Place stationary arm of goniometer at the level of the dorsal aspect of the wrist and parallel to the anterior longitudinal midline of the arm. (The goniometer will be anterior to the body by the length of the forearm)
2. Movable arm is placed across the dorsum of the wrist between the styloid processes of the radius and ulna.

Elbow should be kept close to thorax. Do not allow abduction or rotation at the shoulder joint or lateral flexion of the trunk to the opposite side.

PRIME MOVERS

MUSCLE	ORIGIN	INSERTION
Biceps brachii N: Musculocutaneous (C5, 6)	*Short head:* a. Flattened tendon from apex of coracoid process of scapula *Long head:* a. Tendon from supraglenoid tuberosity	a. Posterior portion of tuberosity of radius and aponeurosis m. bicipitis brachii
Supinator N: Radial (deep branch) (C6)	a. Lateral epicondyle of humerus b. Ridge and depression on ulna distal to radial notch c. Annular ligament and radial collateral ligament of elbow joint	a. Muscle winds around radius to insert into dorsal and lateral surfaces of body of radius between oblique line and head of bone

Accessory Muscle
Brachioradialis

FOREARM SUPINATION

NORMAL AND GOOD

Sitting with arm at side (elbow flexed to 90 degrees to prevent rotation at the shoulder) and forearm pronated. Muscles of wrist and fingers are relaxed.

Stabilize arm.

Patient supinates forearm.

Resistance is given on dorsal surface of distal end of radius with counterpressure against the ventral surface of the ulna.

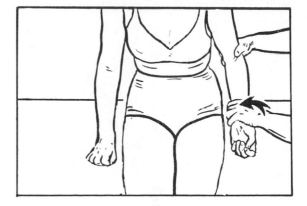

FAIR AND POOR

Sitting with arm at side, elbow flexed to 90 degrees, forearm pronated and supported by examiner. Muscles of wrist and fingers are relaxed.

Patient supinates forearm through full range of motion for Fair grade and through partial range for Poor grade.

Alternate test for Poor grade: Sitting, shoulder flexed to 90 degrees. Elbow flexed and resting on table with forearm perpendicular to table and pronated. Supinate forearm through range of motion.

TRACE AND ZERO

Supinator muscle may be palpated below the head of the radius on dorsal aspect of forearm. Keep wrist flexed with wrist extensors relaxed to differentiate Supinator from extensors.

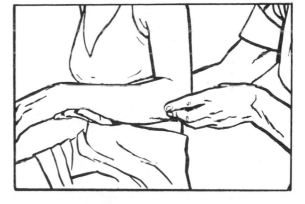

Patient should not be allowed to laterally rotate arm and move elbow across thorax as forearm is supinated. As a result of this movement the forearm may appear to be supinated, but range of motion is incomplete. This motion may "roll" the forearm into supination without a muscular contraction taking place.

FOREARM PRONATION

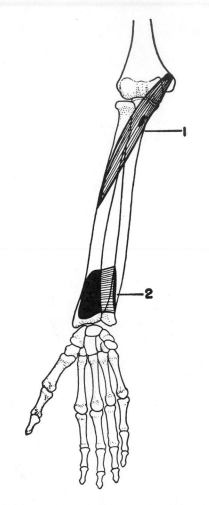

Range of Motion:

0 to 80–90 degrees

Factors Limiting Motion:

1. Tension of dorsal radioulnar, ulnar collateral and dorsal radiocarpal ligaments
2. Tension of lowest fibers of interosseous membrane

Measurement:

Sitting, elbow flexed to 90 degrees, arm against thorax. Forearm supinated.

1. Place stationary arm of goniometer at the level of the ventral aspect of the wrist and parallel to the anterior longitudinal midline of the arm.
2. Movable arm is placed on the widest and most flattened area proximal to the wrist.

Elbow should be kept at the lateral side of the thorax. Do not allow lateral flexion of the trunk to the same side as the extremity being measured.

Palmar View of Forearm and Hand
1. *Pronator teres*
2. *Pronator quadratus*

PRIME MOVERS

Muscle	Origin	Insertion
Pronator teres N: Median (C6, 7)	*Humeral head:* a. Area proximal to medial epicondyle of humerus b. Common tendon of flexor muscle group *Ulnar head:* a. Medial side of coronoid process of ulna (joins humeral head at acute angle)	a. Rough impression on lateral surface of radius at middle of body
Pronator quadratus N: Median (Palmar interosseous branch) (C8, T1)	a. Palmar surface of lower fourth of ulna	a. Distal fourth of lateral border and palmar surface of radius b. Deeper fibers into triangular area proximal to ulnar notch of radius

Accessory Muscle
Flexor carpi radialis

FOREARM PRONATION

NORMAL AND GOOD

Sitting with arm at side, elbow flexed to 90 degrees to prevent rotation at the shoulder and forearm supinated. Muscles of wrist and fingers are relaxed.

Stabilize arm.

Patient pronates forearm through range of motion.

Resistance is given on palmar surface of distal end of radius with counterpressure against the dorsal surface of the ulna for derotation.

Note: If sitting position is contraindicated, all tests can be given in the supine position. Patient completes range of motion with slight resistance for a Fair grade and range only for Poor.

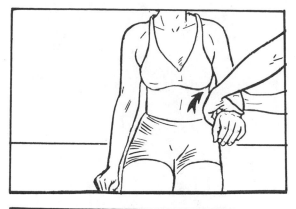

FAIR AND POOR

Sitting with arm at side, elbow flexed to 90 degrees, forearm supinated and supported by examiner. Muscles of wrist and fingers are relaxed.

Patient pronates forearm through full range of motion for Fair grade and through partial range for Poor grade.

See Alternate test for Poor grade under Forearm Supination.

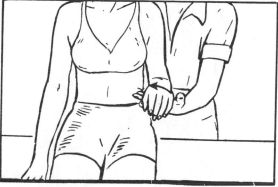

TRACE AND ZERO

Sitting.

Examiner palpates fibers of Pronator teres on upper third of palmar surface of forearm on a diagonal line from medial condyle of humerus to lateral border of radius.

Patient should not be allowed to medially rotate or abduct shoulder during pronation. This movement makes the range of motion in pronation appear complete and allows forearm to roll into pronated position.

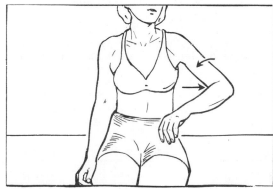

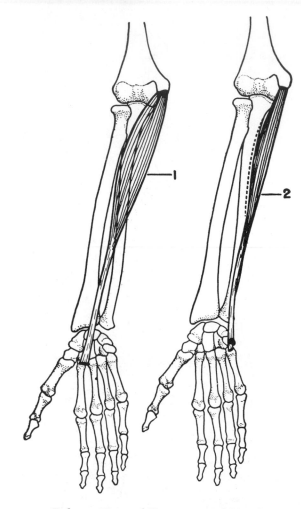

Range of Motion:

0 to 80–90 degrees

Factors Limiting Motion:

Tension of dorsal radiocarpal ligament

Measurement:

Sitting, forearm resting on table in midposition between pronation and supination. Fingers loosely extended. Wrist flexed.

1. Place stationary arm of goniometer on longitudinal midline of dorsal surface of forearm.
2. Movable arm is placed on dorsal surface of third metacarpal.

Do not allow abduction or adduction of the wrist.

Abduction (radial flexion or deviation) and adduction (ulnar flexion or deviation) of wrist can be measured with the same placement of the goniometer.

Abduction: 0 to 15–25 degrees
Adduction: 0 to 30–40 degrees

Palmar View of Forearm and Hand
1. *Flexor carpi radialis*
2. *Flexor carpi ulnaris*

PRIME MOVERS

MUSCLE	ORIGIN	INSERTION
Flexor carpi radialis N: Median (C6, 7)	a. Medial epicondyle of humerus by common tendon	a. Base of second metacarpal bone on palmar surface b. May send slip to base of third metacarpal bone
Flexor carpi ulnaris N: Ulnar (C8, T1)	a. Medial epicondyle of humerus by common tendon (humeral head) b. Medial margin of olecranon process and upper two thirds of dorsal border of ulna (ulnar head)	a. Pisiform bone b. Prolongations to hamate and base of fifth metacarpal bone
Accessory Muscle Palmaris longus		

NORMAL AND GOOD

Forearm in supination with dorsal surface resting on table. (In illustration, forearm is partially pronated to show tendon of Flexor carpi radialis.) Muscles of thumb and fingers relaxed.

Stabilize forearm.

Patient flexes wrist.

To test Flexor carpi radialis, resistance is given at base of second metacarpal bone in direction of extension and ulnar deviation.

To test Flexor carpi ulnaris, resistance is given at base of fifth metacarpal bone in direction of extension and radial deviation. (Not illustrated.)

Note: Do not allow "curling" of fingers prior to flexion or during resistance to prevent substitution by the finger flexors.

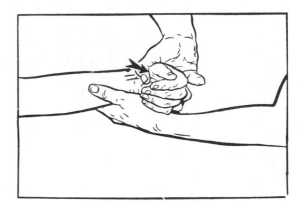

FAIR

Forearm in supination, muscles of thumb and fingers relaxed.

Stabilize forearm.

Patient flexes wrist with radial deviation or ulnar deviation.

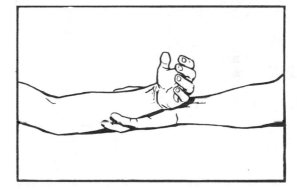

POOR

Forearm and hand on table, forearm in midposition, hand resting on medial border.

Stabilize forearm.

Patient flexes wrist, sliding hand along table. Deviation should be observed and muscles graded accordingly.

A grade of Poor can be given for completing full range of motion.

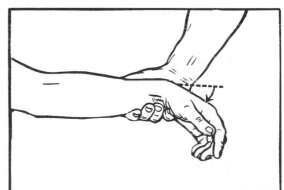

TRACE AND ZERO

Examiner palpates tendon of Flexor carpi radialis on lateral palmar aspect of wrist and tendon of Flexor carpi ulnaris on medial palmar surface as patient attempts to flex the wrist.

WRIST EXTENSION

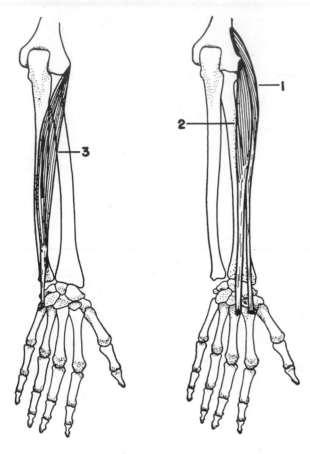

Range of Motion:

0 to 75–85 degrees

Factors Limiting Motion:

Tension of palmar radiocarpal ligament

Measurement:

Sitting, forearm resting on table in midposition between pronation and supination. Fingers loosely flexed. Wrist extended.

1. Place stationary arm of goniometer on longitudinal midline of volar surface of forearm.

2. Movable arm is placed parallel to the volar surface of the third metacarpal. (The goniometer will extend between the third and fourth fingers.)

Do not allow abduction or adduction of the wrist. (See Wrist Flexion.)

Dorsal View of Forearm and Hand
1. *Extensor carpi radialis longus*
2. *Extensor carpi radialis brevis*
3. *Extensor carpi ulnaris*

PRIME MOVERS

Muscle	Origin	Insertion
Extensor carpi radialis longus N: Radial (C6, 7)	a. Distal third of lateral supra-condylar ridge of humerus b. Some fibers from common extensor tendon of lateral epicondyle of humerus	a. Dorsal surface of base of second metacarpal bone on radial side
Extensor carpi radialis brevis N: Radial (C6, 7)	a. Lateral epicondyle of humerus by common tendon b. Radial collateral ligament of elbow joint	a. Dorsal surface of base of third metacarpal bone on radial side
Extensor carpi ulnaris N: Deep radial (C6, 7, 8)	a. Lateral epicondyle of humerus by common extensor tendon b. Aponeurosis from dorsal border of ulna	a. Tubercle on ulnar side of base of fifth metacarpal bone

NORMAL AND GOOD

Forearm in pronation, muscles of thumb and fingers relaxed.

Stabilize forearm.

Patient extends wrist.

To test Extensor carpi radialis longus and brevis, resistance is given on dorsal surface of second and third metacarpal bones in direction of flexion and ulnar deviation.

To test Extensor carpi ulnaris, resistance is given on dorsal surface of fifth metacarpal bone in direction of flexion and radial deviation.

Note: Do not allow extension of fingers during the movement of resistance to prevent substitution by the finger extensors.

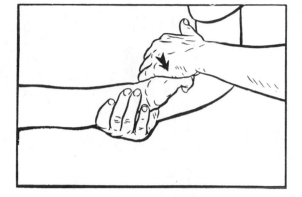

FAIR

Forearm in pronation, muscles of thumb and fingers relaxed.

Stabilize forearm.

Patient extends wrist with radial deviation or ulnar deviation.

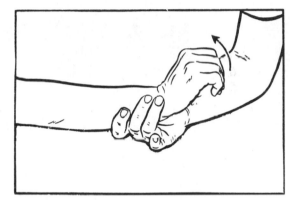

POOR

Forearm and hand on table, forearm in mid-position, hand resting on ulnar border.

Stabilize forearm.

Patient extends wrist, sliding hand along table through range of motion.

Deviation should be observed and muscles graded accordingly.

A grade of Poor can be given for completing full range of motion.

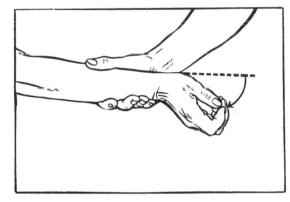

TRACE AND ZERO

Tendons of radial wrist extensors may be found on lateral dorsal surface of wrist in line with second and third metacarpal bones and ulnar on medial dorsal surface proximal to fifth metacarpal bone as patient attempts to extend wrist.

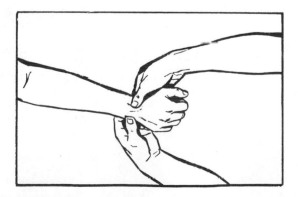

FLEXION OF METACARPOPHALANGEAL JOINTS OF FINGERS

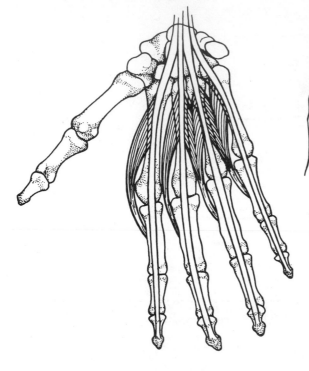

Palmar View
Lumbricals

Range of Motion:

0 to 85–105 degrees

Factors Limiting Motion:

Tension of expansions of extensor tendons of fingers.

Measurement:

Sitting, forearm resting on table in midposition between pronation and supination. Fingers extended, metacarpophalangeal joints flexed.

1. Place stationary arm of goniometer on the dorsal surface of the metacarpal of each joint to be measured.
2. Movable arm is placed on dorsal surface of proximal phalanx.

Range of motion is least in index finger and increases slightly in each finger in turn.
Use a small goniometer.

PRIME MOVERS

MUSCLE	ORIGIN	INSERTION
Lumbricals First and second lumbricals N: Median (C6, 7) Third and fourth lumbricals N: Ulnar (deep palmar branch C8)	a. Arise from tendons of Flexor digitorum profundus muscle *First and second:* Radial sides and palmar surfaces of tendons of index and middle fingers *Third:* Adjacent sides of tendons of middle and ring fingers *Fourth:* Adjacent sides of tendons of ring and little fingers	a. Pass to radial side of corresponding fingers to insert opposite metacarpophalangeal joints into tendinous expansions of Extensor digitorum communis, covering dorsal aspect of fingers
Dorsal interossei N: Ulnar (deep palmar branch C8, T1)	a. Each by 2 heads from adjacent sides of metacarpal bones between which it lies	a. Base of proximal phalanges of 4 fingers: first and second into radial side of index and middle fingers; third and fourth into ulnar side of middle and ring fingers

(Continued on Page 131)

FLEXION OF METACARPOPHALANGEAL JOINTS OF FINGERS

NORMAL, GOOD AND FAIR

Forearm supinated. Stabilize metacarpals.

Patient flexes fingers at metacarpophalangeal joints, keeping interphalangeal joints extended.

Resistance is given on palmar surface of proximal row of phalanges.

Extension of the middle and distal interphalangeal joints with the metacarpophalangeal joints in flexion should be tested concurrently. Extension of these joints is part of the primary action of the Lumbricals.

Patient flexes fingers through the full range of motion for Fair grade.

Note: Resistance should be given to each finger separately as Lumbricals are uneven in strength and have divided innervation.

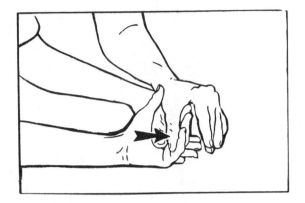

POOR

Forearm and wrist in midposition. Stabilize metacarpals.

Patient flexes metacarpophalangeal joints through full range of motion with interphalangeal joints extended for Poor grade.

TRACE AND ZERO

Contraction of Lumbricals may be detected by light pressure against palmar surface of proximal phalanges as patient attempts to flex at metacarpophalangeal joints.

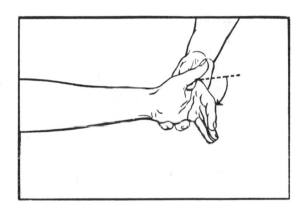

PRIME MOVERS *(Continued)*

MUSCLE	ORIGIN	INSERTION
Palmar interossei N: Ulnar (deep palmar branch C8, T1)	a. Entire length of second, fourth, and fifth metacarpal bones on palmar surface	a. Side of base of proximal phalanx of corresponding finger; first into ulnar side of index finger; second and third into radial side of ring and little fingers b. Into aponeurotic expansion of Extensor digitorum communis tendons of same fingers

Accessory Muscles
Flexor digiti minimi brevis
Flexor digitorum superficialis
Flexor digitorum profundus

FLEXION OF PROXIMAL AND DISTAL INTERPHALANGEAL JOINTS OF FINGERS

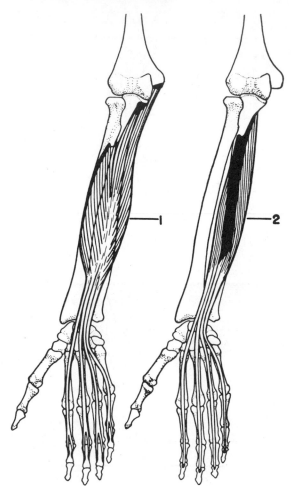

Palmar View of Forearm and Hand
1. *Flexor digitorum superficialis*
2. *Flexor digitorum profundus*

Range of Motion:

Proximal—0 to 110–120 degrees
Distal—0 to 80–90 degrees

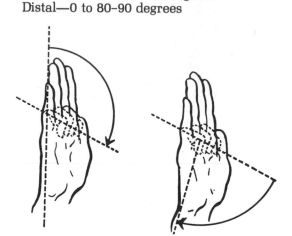

Factors Limiting Motion:

Tension of expansions of Extensor digitorum tendons (dorsal ligaments)

Measurement:

Sitting, elbow or forearm resting on table, wrist partially extended to lessen tension in the long extensors of the fingers. Proximal interphalangeal joints are first flexed for measurement, then distal interphalangeal joints are flexed and measured.

The midlines of the dorsal surface of the proximal, middle, and distal phalanges are used for placement of the goniometer.

Flexion of Proximal Interphalangeal Joints of Fingers

PRIME MOVERS

MUSCLE	ORIGIN	INSERTION
Flexor digitorum superficialis N: Median (C7, 8, T1)	a. Medial epicondyle of humerus by common flexor tendon (humeral head) b. Medial side of coronoid process of ulna (ulnar head) c. Oblique line of radius from radial tuberosity to insertion of Pronator teres (radial head)	*Tendon divides:* a. Superficial part into middle and ring fingers (insertions into sides of second phalanges) b. Deep part into index and little fingers (insertions into sides of second phalanges)

(Continued on Page 133)

FLEXION OF PROXIMAL AND DISTAL INTERPHALANGEAL JOINTS OF FINGERS

Flexion of Proximal Interphalangeal Joints of Fingers (Flexor Digitorum Superficialis)

NORMAL AND GOOD

Forearm supinated, wrist in midposition and fingers extended.

Stabilize proximal phalanx of finger.

Patient flexes middle phalanx, and resistance is given on palmar surface.

FAIR AND POOR

Patient flexes middle phalanx through full range of motion for Fair grade and through partial range for Poor grade.

Flexion of Distal Interphalangeal Joints of Fingers (Flexor Digitorum Profundus)

NORMAL AND GOOD

Forearm supinated, wrist in midposition with proximal interphalangeal joint in extension.

Stabilize middle phalanx of finger.

Patient flexes distal phalanx. Resistance is given on palmar surface of distal phalanx of finger.

FAIR AND POOR

Patient flexes distal phalanx through full range of motion for fair grade and through partial range for poor grade.

TRACE AND ZERO

Flexor digitorum profundus may be palpated on the palmar surface of the middle phalanx.

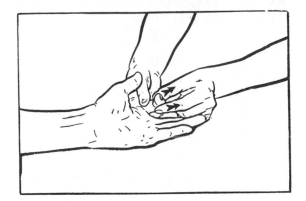

Note: Keep fingers adjacent to the one being tested in extension at the metacarpophalangeal joints to prevent substitution by the Flexor digitorum profundus.

Flexion of the wrist allows "slack" in the long finger flexors and increases tension in the long finger extensors that inhibit completion of the range of motion in flexion.

Extension of the wrist causes tension in the long finger flexors that results in passive flexion at the interphalangeal joints.

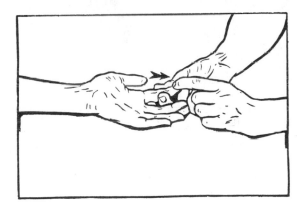

Flexion of Distal Interphalangeal Joints of Fingers

PRIME MOVERS

MUSCLE	ORIGIN	INSERTION
Flexor digitorum profundus N: Ulnar (C8, T1) and median (palmar interosseus branch C8, T1)	a. Proximal three fourths of volar and medial surfaces of ulna b. Aponeurosis from upper three fourths of dorsal border of ulna c. Medial side of coronoid process	a. Bases of distal phalanges of 4 fingers (tendons pass through those of Flexor digitorum superficialis)

EXTENSION OF METACARPOPHALANGEAL JOINTS OF FINGERS

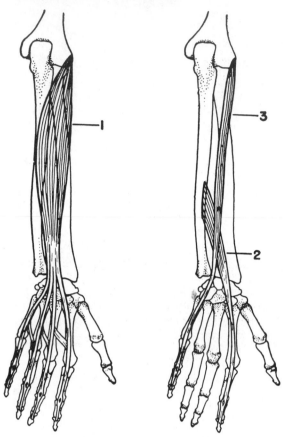

Range of Motion:

0 to 20–30 degrees

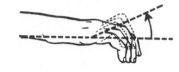

Factors Limiting Motion:

1. Tension of palmar and collateral ligaments
2. Tension of flexor muscles of fingers

Measurement:

Sitting, forearm resting on table in midposition between pronation and supination. Fingers partially flexed at the interphalangeal joints. Metacarpophalangeal joints extended.

1. Place stationary arm of goniometer on the dorsal surface of the metacarpal of each joint to be measured.
2. Movable arm is placed on the dorsal surface of the proximal phalanx.

Dorsal View of Forearm and Hand
1. *Extensor digitorum*
2. *Extensor indicis*
3. *Extensor digiti minimi*

PRIME MOVERS

Muscle	Origin	Insertion
Extensor digitorum N: Deep radial (C6, 7, 8)	a. Lateral epicondyle of humerus by common tendon	a. Four tendons into base of second and third phalanges of fingers (Opposite metacarpophalangeal joints, tendons are bound by fasciculi to collateral ligaments)
Extensor indicis N: Deep radial (C6, 7, 8)	a. Dorsal surface of body of ulna below origin of Extensor pollicis longus	a. Joins ulnar side of tendon of Extensor digitorum communis, which goes to index finger; terminates in extensor expansion
Extensor digiti minimi N: Deep radial (C6, 7, 8)	a. Common extensor tendon from lateral epicondyle of humerus	a. Joins expansion of Extensor digitorum communis tendon on dorsum of the first phalanx of fifth finger

EXTENSION OF METACARPOPHALANGEAL JOINTS OF FINGERS

NORMAL, GOOD AND FAIR

Forearm pronated, wrist in midposition, fingers flexed.

Stabilize metacarpals.

Patient extends metacarpophalangeal joints with interphalangeal joints partially flexed. Resistance is given on dorsal surface of proximal row of phalanges of fingers. Patient extends metacarpophalangeal joints through full range for Fair grade.

Note: Resistance should be given to each finger separately. Extensor indicis assists in extension of index finger, and Extensor digiti minimi assists in extension of fifth finger.

Extension of the wrist beyond the midline allows "slack" in the long finger extensors and tension in the long finger flexors that inhibit completion of range of motion in extension.

Flexion of the wrist causes tension in the long finger extensors that results in passive extension at the metacarpophalangeal joint.

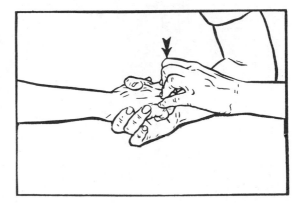

POOR

Forearm and wrist in midposition, fingers flexed.

Stabilize metacarpals.

Patient extends metacarpophalangeal joints through full range of motion for grade of Poor.

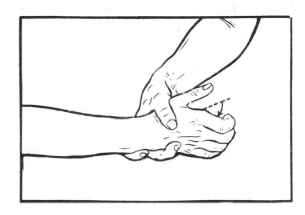

TRACE AND ZERO

The tendons of the finger extensors may easily be located on dorsum of hand where they pass over metacarpals.

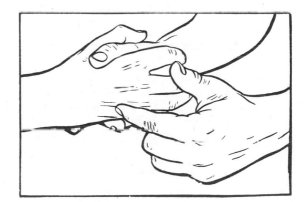

FINGER ABDUCTION

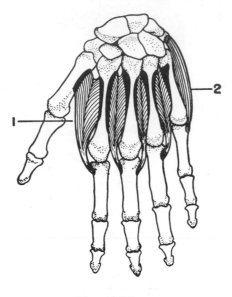

Dorsal View
1. Dorsal interossei
2. Abductor digiti minimi

Range of Motion:

0 to 20–25 degrees

Factors Limiting Motion:

Tension of fascia and skin between fingers

Measurement:

Sitting, forearm resting on table in prona-tion, palm flat. Fingers are abducted. (Middle finger remains straight but can be actively ab-ducted medially or laterally.)

1.   Place stationary arm of goniometer on the dorsal surface of each metacarpal.
2.   Movable arm is placed on proximal phalanx of each finger to be measured.

PRIME MOVERS

Muscle	Origin	Insertion
Dorsal interossei N: Ulnar (deep palmar branch C8, T1)	a. Each by 2 heads from adjacent sides of metacarpal bones be-tween which it lies	a. Base of proximal phalanges of 3 fingers: the first into the radial side of index finger, the second and third into the middle finger, and the fourth into the ulnar side of the ring finger. b. Into aponeurotic expansions of Extensor digitorum tendons of corresponding fingers
Abductor digiti minimi N: Ulnar (C8, T1)	a. Pisiform bone b. Tendon of Flexor carpi ulnaris	*Tendon divides:* a. One part into ulnar side of first phalanx of little finger at base b. Other part into ulnar border of aponeurosis of Extensor digiti minimi muscle

FINGER ABDUCTION

NORMAL AND GOOD

(Test for first and third Dorsal interossei)

Forearm pronated, hand resting on table, fingers in extension and adduction.
  Stabilize metacarpals.
  Patient abducts fingers.
  Resistance is given on radial side of index and ulnar side of middle finger.
  To test individual fingers, resistance is given on proximal phalanx.

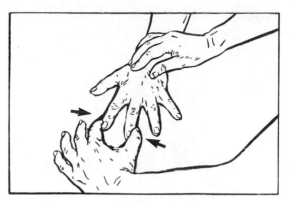

NORMAL AND GOOD

(Test for second and fourth Dorsal interossei and Abductor digiti minimi)

Patient abducts fingers.
  Resistance is given on ulnar side of ring and little fingers and on radial side of middle finger.

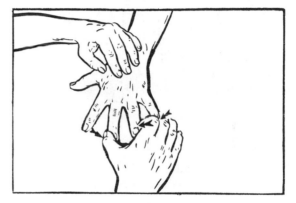

FAIR AND POOR

Forearm pronated, hand resting on table, fingers in extension and adduction.
  Patient abducts fingers through full range of motion for Fair grade and through partial range for Poor grade. Middle finger must be moved in both directions.

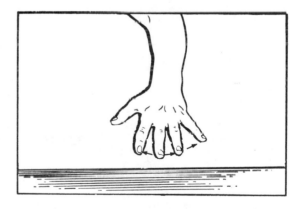

TRACE AND ZERO

The Dorsal interossei lie deep between the metacarpal bones on the dorsum of the hand. The Abductor digiti minimi may be palpated easily on the lateral border of the fifth metacarpal bone. (Palpation of first Dorsal interossei shown in illustration.)

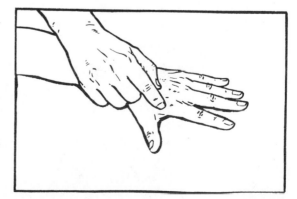

FINGER ADDUCTION

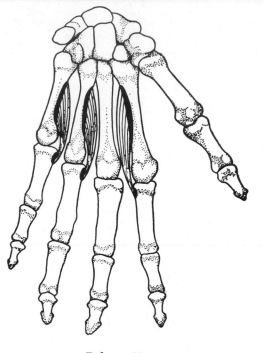

Palmar View
Palmar interossei

Range of Motion:

20–25 to 0 degrees

Factors Limiting Motion:

Contact of fingers

Measurement:

Sitting, forearm resting on table in pronation, palm flat. Fingers are adducted.

Arms of the goniometer are placed in the same positions as for abduction of the fingers.

PRIME MOVER

Muscle	Origin	Insertion
Palmar interossei N: Ulnar (deep palmar branch C8, T1)	a. Entire length of second, fourth, and fifth metacarpal bones on palmar surface	a. Side of base of proximal phalanx of corresponding finger: first into ulnar side of index finger, second and third into radial side of ring and little fingers b. Into aponeurotic expansion of Extensor digitorum tendon of same finger

FINGER ADDUCTION

NORMAL AND GOOD

Forearm pronated, fingers in extension and abduction.

Patient adducts fingers.

Resistance is given in radial direction on proximal phalanx of index finger and in ulnar direction on ring and little fingers.

Test fingers individually.

FAIR AND POOR

Forearm pronated, hand resting on table, fingers in extension and abduction.

Patient adducts fingers through full range of motion for Fair grade and through partial range for Poor grade.

TRACE AND ZERO

Presence of contraction of the Palmar intcrossci may be determined by outward pressure on the index, ring, and little fingers as the patient attempts to adduct.

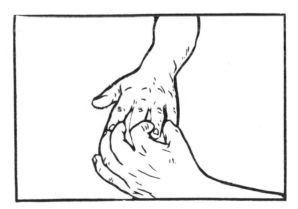

FLEXION OF METACARPOPHALANGEAL AND INTERPHALANGEAL JOINTS OF THUMB

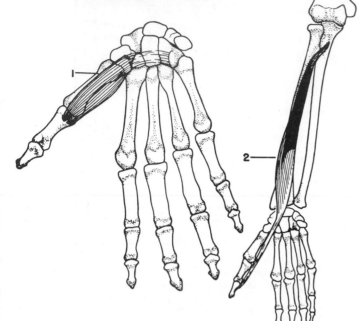

Palmar View of Forearm and Hand
1. *Flexor pollicis brevis*
2. *Flexor pollicis longus*

Range of Motion:

Metacarpophalangeal flexion: 0 to 60–70 degrees

Interphalangeal flexion: 0 to 80–90 degrees

Factors Limiting Motion:

Tension of tendons of extensor muscles of thumb

Measurement:

Sitting, forearm resting on table in midposition between supination and pronation, wrist partially extended to lessen tension in the long extensors of the thumb.

The midline on the dorsal surface of the first metacarpal, the proximal phalanx, and the distal phalanx are used for placement of the goniometer in measuring flexion of the metacarpophalangeal joint and the distal joint of the thumb.

Flexion of Metacarpophalangeal Joint of Thumb

PRIME MOVER

Muscle	Origin	Insertion
Flexor pollicis brevis (lateral portion) N: Median (C6, 7)	*Lateral portion* (superficial): a. Distal border of flexor retinaculum b. Ridge on trapezium bone	a. Base of proximal phalanx of thumb on radial side (sesamoid bone)
(medial portion) N: Ulnar (deep palmar branch C8, T1)	*Medial portion* (deep): a. Ulnar side of first metacarpal bone between Adductor pollicis (oblique) and lateral head of first Dorsal interossei	a. Ulnar side of base of proximal phalanx of thumb with Adductor pollicis (oblique)

Flexion of Interphalangeal Joint of Thumb

PRIME MOVER

Muscle	Origin	Insertion
Flexor pollicis longus N: Median (palmar interosseous branch C8, T1)	a. Volar surface of body of radius from tuberosity to attachment of Pronator quadratus b. Interosseous membrane c. Usually from coronoid process or medial epicondyle of humerus	a. Base of distal phalanx of thumb

FLEXION OF METACARPOPHALANGEAL AND INTERPHALANGEAL JOINTS OF THUMB

Flexion of Metacarpophalangeal Joint of Thumb

NORMAL AND GOOD

Forearm in supination, wrist in midposition. Stabilize first metacarpal.

Patient flexes proximal phalanx of thumb. Distal phalanx remains relaxed.

Resistance is given on palmar surface of proximal phalanx.

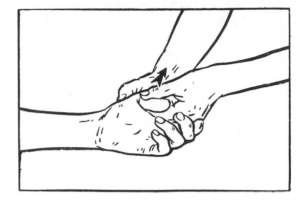

FAIR AND POOR

Patient flexes proximal phalanx of thumb through full range of motion for Fair grade and through partial range for Poor grade.

TRACE AND ZERO

Contraction of Flexor pollicis brevis may be determined by pressure over palmar surface of first metacarpal (medial to Abductor pollicis brevis) as patient attempts flexion.

Flexor pollicis longus may substitute. Do not allow flexion of distal joint of thumb.

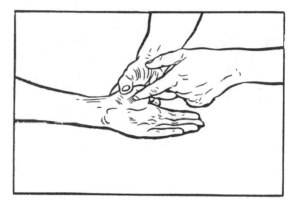

Flexion of Interphalangeal Joint of Thumb

NORMAL AND GOOD

Forearm in supination, wrist in midposition. Stabilize proximal phalanx of thumb.

Patient flexes distal phalanx (motion takes place in plane of palm). Resistance is given on palmar surface of distal phalanx of thumb.

FAIR AND POOR

Patient flexes distal phalanx through full range of motion for Fair grade and through partial range for Poor grade.

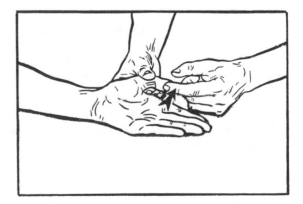

TRACE AND ZERO

The tendon of Flexor pollicis longus may be found on palmar surface of the proximal phalanx of the thumb (as shown in the illustration.)

EXTENSION OF METACARPOPHALANGEAL AND INTERPHALANGEAL JOINTS OF THUMB

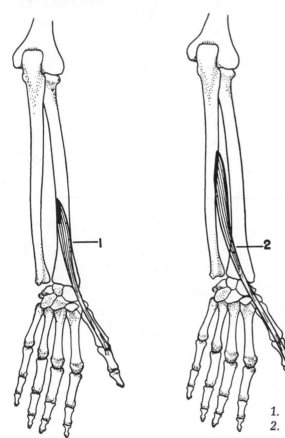

Range of Motion:

Metacarpophalangeal extension: 60–70 to 0 degrees

Interphalangeal extension: 80–90 to 0 degrees

Factors Limiting Motion:

Tension of palmar and collateral ligaments of thumb

Measurement:

Sitting, forearm in midposition, wrist partially flexed to lessen tension in the long flexor of the thumb.

Placement of goniometer is the same as for Flexion of the Metacarpophalangeal and Interphalangeal Joints of the Thumb.

Dorsal View
1. *Extensor pollicis brevis*
2. *Extensor pollicis longus*

Extension of Metacarpophalangeal Joint of Thumb

PRIME MOVER

Muscle	Origin	Insertion
Extensor pollicis brevis N: Deep radial (C6, 7)	a. Dorsal surface of body of radius distal to Abductor pollicis longus muscle b. Interosseous membrane	a. Base of proximal phalanx of thumb on dorsal aspect

Extension of Interphalangeal Joint of Thumb

PRIME MOVER

Muscle	Origin	Insertion
Extensor pollicis longus N: Deep radial (C6, 7, 8)	a. Lateral part of middle third of body of ulna on dorsal surface below Abductor pollicis longus muscle b. Interosseous membrane	a. Base of distal phalanx of thumb on dorsal aspect

EXTENSION OF METACARPOPHALANGEAL AND INTERPHALANGEAL JOINTS OF THUMB

Extension of Metacarpophalangeal Joint of Thumb

NORMAL AND GOOD

Forearm and wrist in midposition.
Stabilize first metacarpal.
Patient extends proximal phalanx of thumb.
Resistance is given on dorsal surface of proximal phalanx.

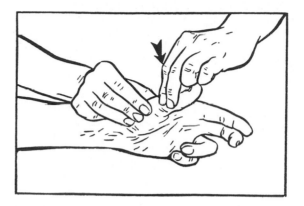

FAIR AND POOR

Patient extends proximal phalanx of thumb through full range of motion for Fair and through partial range for Poor. (Not illustrated.)

TRACE AND ZERO

Tendon of Extensor pollicis brevis may be palpated at the base of the first metacarpal between the tendons of the Abductor and Extensor pollicis longus.

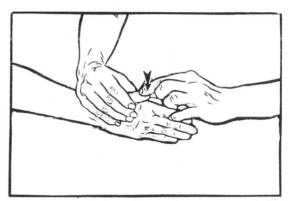

Extension of Interphalangeal Joint of Thumb

NORMAL AND GOOD

Forearm and wrist in midposition.
Stabilize proximal phalanx of thumb.
Patient extends distal phalanx (motion takes place in plane of palm). Resistance is given on dorsal surface of distal phalanx of thumb.

FAIR AND POOR

Patient extends distal phalanx of thumb through full range of motion for Fair and through partial range for Poor. (Not illustrated.)

TRACE AND ZERO

Tendon of Extensor pollicis longus may be palpated on dorsal surface of hand between head of first metacarpal and base of second. It may also be found on dorsal surface of proximal phalanx.

THUMB ABDUCTION

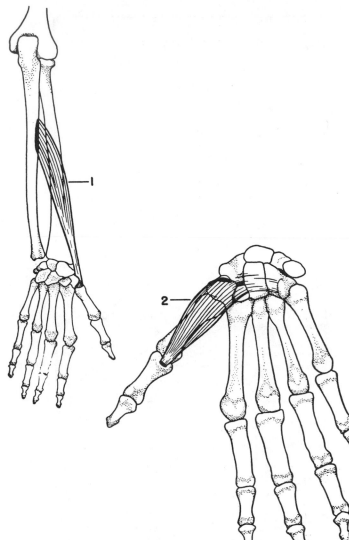

Range of Motion:

0 to 70–80 degrees
(Motion takes place primarily at carpometacarpal joint)

Factors Limiting Motion:

1. Tension of skin between thumb and index finger
2. Tension of first Dorsal interossei muscle

Measurement:

Sitting, forearm resting on table in midposition between pronation and supination. Thumb is abducted.

1. Place stationary arm of goniometer on lateral border of second metacarpal.
2. Movable arm is placed on the medial border of the dorsal surface of the first metacarpal.

Dorsal View of Forearm and Hand
1. *Abductor pollicis longus*

Palmar View of Hand
2. *Abductor pollicis brevis*

PRIME MOVERS

MUSCLE	ORIGIN	INSERTION
Abductor pollicis longus N: Deep radial (C6, 7)	a. Lateral part of dorsal surface of body of ulna below Anconeus muscle b. Middle third of dorsal surface of body of radius	a. Radial side of base of first metacarpal bone
Abductor pollicis brevis N: Median (C6, 7)	a. Tuberosity of scaphoid bone b. Ridge of trapezium c. Transverse carpal ligament	a. Radial side of base of first phalanx of thumb b. Capsule of first metacarpophalangeal joint
Accessory Muscle Palmaris longus		

NORMAL AND GOOD

Forearm supinated, wrist in midposition.

Stabilize medial four metacarpals and wrist.

Patient raises thumb vertically through range of abduction.

Resistance is given on lateral border of proximal phalanx of the thumb for the Abductor pollicis brevis and the distal end of the metacarpal for the Abductor pollicis longus.

If the Abductor pollicis longus is stronger than the brevis, thumb will deviate toward radial side of hand.

If the Abductor pollicis brevis is stronger, deviation will be toward ulnar side.

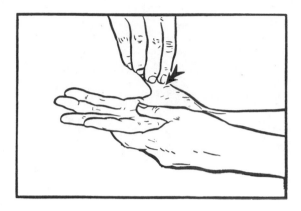

FAIR AND POOR

Forearm supinated, wrist in midposition. (Forearm partially pronated in illustration to show range of motion.)

Stabilize metacarpals and wrist.

Patient abducts thumb through full range of motion for Fair grade and through partial range for Poor grade.

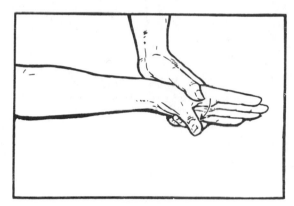

TRACE AND ZERO

The Abductor pollicis brevis fibers may easily be palpated on thenar eminence lateral to the Flexor pollicis brevis. The tendon of the Abductor pollicis longus may be palpated near its insertion on the radial side of the base of the first metacarpal bone and is the most lateral tendon at the wrist. (Latter not illustrated.)

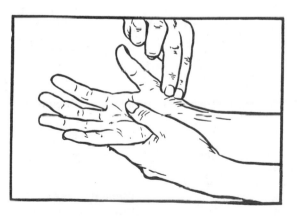

THUMB ADDUCTION

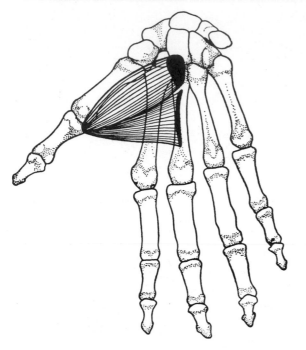

Palmar View
Adductor pollicis: transverse head and oblique head

Range of Motion:

70–80 to 0 degrees
(Motion takes place primarily at carpometacarpal joint)

Factors Limiting Motion:

Contact of thumb with second metacarpal bone

Measurement:

Sitting, forearm resting on table in midposition between pronation and supination. Thumb is adducted.

The placement of the goniometer is the same as for Thumb Abduction.

PRIME MOVERS

MUSCLE	ORIGIN	INSERTION
Adductor pollicis, oblique head N: Ulnar (Deep palmar branch C8, T1)	a. Capitate bone b. Base of second and third metacarpal bones on palmar surface c. Intercarpal ligaments	a. Unites with tendon of Flexor pollicis brevis and Adductor pollicis (transverse) to insert into base of proximal phalanx of thumb on ulnar side (sesamoid bone) b. Fasciculus to lateral portion of Flexor pollicis brevis and Abductor pollicis brevis
Adductor pollicis, transverse head N: Ulnar (Deep palmar branch C8, T1)	a. Distal two thirds of palmar surface of third metacarpal bone	a. Medial part of Flexor pollicis brevis and Adductor pollicis (oblique) into base of proximal phalanx of thumb on ulnar side

THUMB ADDUCTION

NORMAL AND GOOD

Forearm pronated, wrist in midposition.
Stabilize medial four metacarpals.
Patient adducts thumb.
Resistance is given on medial border of proximal phalanx.

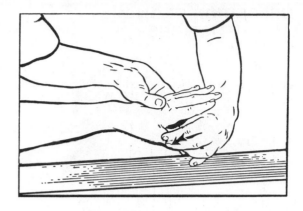

FAIR AND POOR

Forearm pronated, wrist in midposition. (Forearm partially supinated in illustration to show range of motion.)
Stabilize metacarpals.
Patient adducts thumb through full range of motion for Fair grade and through partial range for Poor grade.

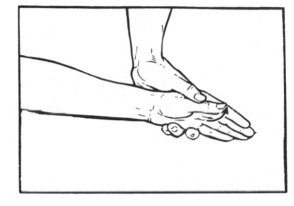

TRACE AND ZERO

Muscle fibers may be palpated between first Dorsal interossei muscle and first metacarpal bone.

Flexor pollicis longus and Flexor pollicis brevis may help pull thumb toward palm. These muscles should remain relaxed during test.

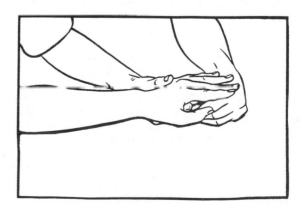

OPPOSITION OF THUMB AND FIFTH FINGER

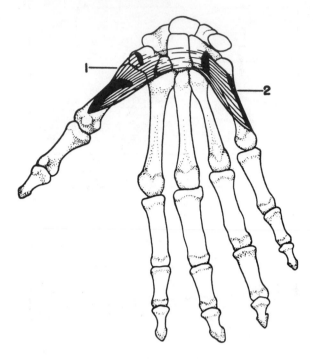

Palmar View
1. Opponens pollicis
2. Opponens digiti minimi

Range of Motion:

Pad of thumb should be brought flat against that of little finger with apposition of first and fifth metacarpal bones

Factors Limiting Motion:

1. Tension of transverse metacarpal ligament (limits motion of fifth metacarpal)
2. Tension of extensor tendons of first and fifth digits

Fixation:

Weight of forearm and hand

PRIME MOVERS

MUSCLE	ORIGIN	INSERTION
Opponens pollicis N: Median (C6, 7)	a. Ridge on trapezium bone b. Flexor retinaculum	a. Entire length of first meta carpal bone on radial side
Opponens digiti minimi N: Ulnar (C8, T1)	a. Convexity of hamulus of hamate bone b. Flexor retinaculum	a. Entire length of fifth metacarpal bone on ulnar side

Accessory Muscles

From the normal rest position the motion of thumb abduction must precede that of opposition. Therefore the Abductor pollicis longus and brevis muscles are in that sense accessory to the total motion.

OPPOSITION OF THUMB AND FIFTH FINGER

NORMAL AND GOOD

Forearm supinated, wrist in midposition.

Patient brings palmar surfaces of distal phalanges of thumb and fifth finger together.

The first and fifth metacarpals rotate toward the midline of the hand. The movement cannot be carried out by muscles other than the two opponens.

Resistance is given on distal end of first and fifth metacarpals on palmar surface with derotating pressure. The two muscles are graded separately.

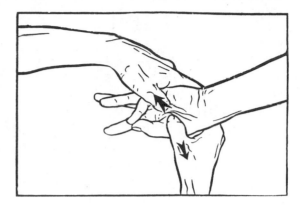

FAIR AND POOR

Patient moves thumb and fifth finger through full range of opposition for Fair grade and through partial range for Poor grade.

The two muscles are graded separately.

TRACE AND ZERO

The edge of the Opponens pollicis will be found lateral to the Abductor pollicis brevis and the insertion of the Opponens digiti minimi on the radial side of the fifth metacarpal.

The muscles of opposition lie under the short abductors and flexors. They are difficult to palpate unless the overlying superficial muscles are nonfunctioning.

Flexor pollicis longus and Flexor pollicis brevis may draw thumb across hand toward fifth finger. This motion takes place in same plane as palm and should be distinguished from true opposition.

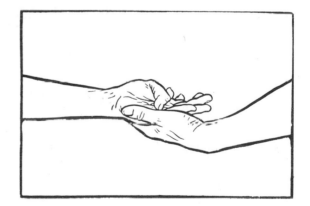

INNERVATION OF THE MUSCLES OF MASTICATION AND OF FACIAL EXPRESSION

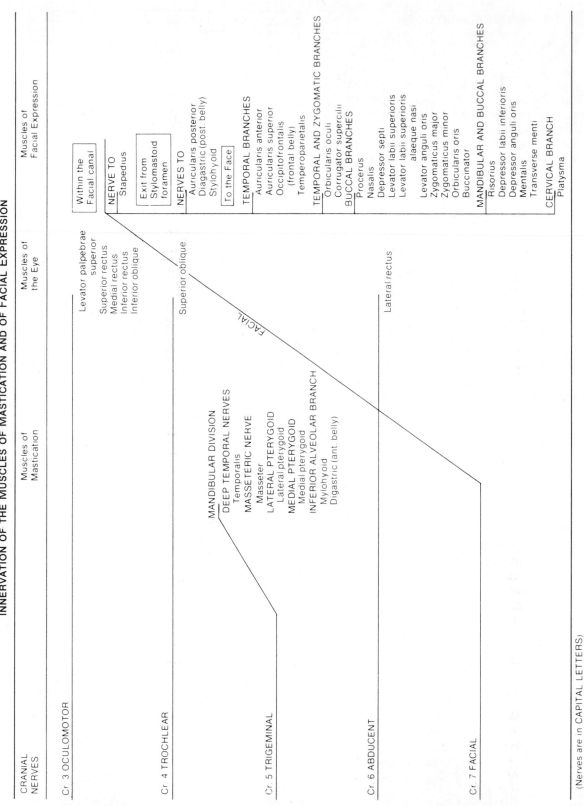

CRANIAL NERVES	Muscles of Mastication	Muscles of the Eye	Muscles of Facial Expression
Cr. 3 OCULOMOTOR		Levator palpebrae superior Superior rectus Medial rectus Inferior rectus Inferior oblique	
Cr. 4 TROCHLEAR		Superior oblique	

Cr. 5 TRIGEMINAL	MANDIBULAR DIVISION
	DEEP TEMPORAL NERVES
	  Temporalis
	MASSETERIC NERVE
	  Masseter
	LATERAL PTERYGOID
	  Lateral pterygoid
	MEDIAL PTERYGOID
	  Medial pterygoid
	INFERIOR ALVEOLAR BRANCH
	  Mylohyoid
	  Digastric (ant. belly)		

Cr. 6 ABDUCENT		Lateral rectus	

Cr. 7 FACIAL			FACIAL
			
			Within the Facial canal
			NERVE TO
			  Stapedius
			Exit from Stylomastoid foramen
			NERVES TO
			  Auricularis posterior
			  Diagastric (post. belly)
			  Stylohyoid
			To the Face
			TEMPORAL BRANCHES
			  Auricularis anterior
			  Auricularis superior
			  Occipitofrontalis (frontal belly)
			  Temperoparietalis
			TEMPORAL AND ZYGOMATIC BRANCHES
			  Orbicularis oculi
			  Corrugator supercilii
			BUCCAL BRANCHES
			  Procerus
			  Nasalis
			  Depressor septi
			  Levator labii superioris
			  Levator labii superioris alaeque nasi
			  Levator anguli oris
			  Zygomaticus major
			  Zygomaticus minor
			  Orbicularis oris
			  Buccinator
			MANDIBULAR AND BUCCAL BRANCHES
			  Risorius
			  Depressor labii inferioris
			  Depressor anguli oris
			  Mentalis
			  Transverse menti
			CERVICAL BRANCH
			  Platysma

(Nerves are in CAPITAL LETTERS)

FACE

In the testing of the face muscles, positioning is not a factor, and, with the exception of the muscles of mastication, only very fine movements are involved. Grades that may be used are: *Zero*, if no contraction can be elicited; *Trace*, for minimal muscle contraction; *Fair*, for performance of the movement with difficulty; and *Normal*, for completion of the movement with ease and control. Resistance may be given in the tests for the muscles of mastication.

MUSCLES OF THE FOREHEAD
AND NOSE

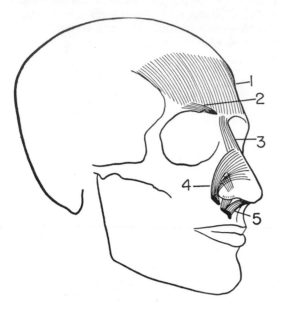

1. *Occipitofrontalis (frontal belly)*
2. *Corrugator supercilii*
3. *Procerus*
4. *Nasalis*
5. *Depressor septi*

PRIME MOVERS

MUSCLE	ORIGIN	INSERTION
Occipitofrontalis (frontal belly) N: Temporal branches of facial (Cr. 7)	Galea aponeurotica midway between coronal suture and orbital arch	Fibers are continuous medially with those of Procerus; intermediate fibers blend with Corrugator and Orbicularis oculi
Corrugator supercilii N: Temporal and zygomatic branches of facial (Cr. 7)	Medial end of superciliary arch	Deep surface of skin above middle of orbital arch
Procerus N: Buccal branches of facial (Cr. 7)	Fascia covering lower part of nasal bone and upper part of lateral nasal cartilage	Skin over lower forehead, between eyebrows
Nasalis N: Buccal branches of facial (Cr. 7)	Transverse part (compressor): Maxilla, above and lateral to incisive fossa Alar part (dilator): Greater alar cartilage	Thin aponeurosis continuous with muscle of opposite side Integument at point of nose
Depressor septi N: Buccal branches of facial (Cr. 7)	Incisive fossa of maxilla	Septum and back part of ala of nose

Muscles of the Forehead and Nose

OCCIPITOFRONTALIS

(frontal belly)

Patient raises eyebrows, forming horizontal wrinkles in forehead (expression of surprise).

CORRUGATOR SUPERCILII

Patient draws eyebrows medially and downward, forming vertical wrinkles between brows (frowning).

PROCERUS

Patient lifts lateral borders of nostrils, forming diagonal wrinkles along bridge of nose (expression of distaste).

NASALIS

Patient dilates nostrils (alar part of Nasalis) followed by compression (transverse portion).

MUSCLES OF THE EYE

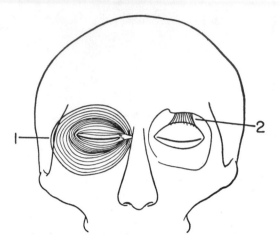

It is probable that no muscle of the eye acts independently. However, the primary action of the Lateral and Medial recti is to abduct or adduct the eye (center of cornea) around a vertical axis; the Superior oblique depresses, and the Inferior oblique elevates the eye (center of cornea) around a lateromedial axis. The eye also rotates around an anteroposterior axis passing through the apex of the cornea (inward and outward movement).

1. *Orbicularis oculi*
2. *Levator palpebrae superioris*

Diagram of eye muscle action.

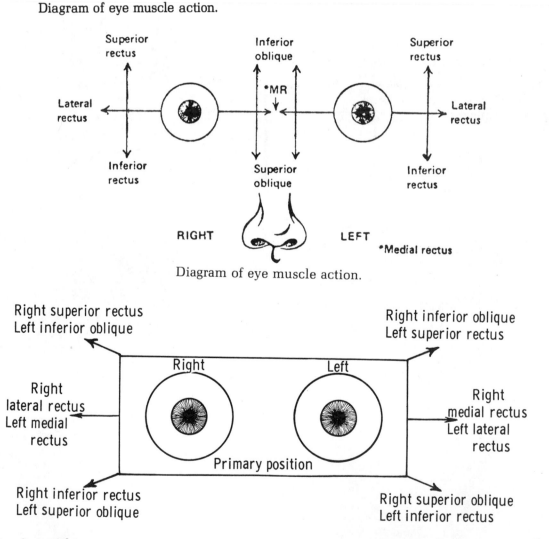

Diagram of eye muscle action.

Muscles used in conjugate ocular movements in the 6 cardinal directions of gaze.

From Chusid, J.G. *Correlative Neuroanatomy and Functional Neurology*, 17th ed. Los Altos, Calif., Lange Medical Publications, 1979, p. 94.

MUSCLES OF THE EYE

ORBICULARIS OCULI

Patient closes eyes tightly.

LEVATOR PALPEBRAE SUPERIORIS

Upper eyelids are lifted completely as eyes are turned upward.

RIGHT SUPERIOR RECTUS AND LEFT INFERIOR OBLIQUE

Patient moves eyes in a direction upward and to the right.

RIGHT SUPERIOR OBLIQUE AND LEFT INFERIOR RECTUS

Patient moves eyes in a direction downward and to the left.

The Medial rectus and Lateral rectus may be tested by movement of the eyes horizontally to the right and left.

PRIME MOVERS

MUSCLE	ORIGIN	INSERTION
Orbicularis oculi N: Temporal and zygomatic branches of facial (Cr. 7)	Orbital part: a. Nasal part of frontal bone b. Frontal process of maxilla in front of lacrimal groove c. Anterior surface and borders of medial palpebral ligament	(Fibers form a complete ellipse without interruption, surrounding circumference of orbit and spreading over temple and downward on cheek)
	Palpebral part: Bifurcation of medial palpebral ligament	Lateral palpebral raphe
	Lacrimal part (tensor tarsi): posterior crest and adjacent part of lacrimal bone	Divides into two slips which insert into superior and inferior tarsi medial to puncta lacrimalia

(Continued on Page 156)

Muscle	Origin	Insertion
Levator palpebrae superioris N: Oculomotor (Cr. 3)	Inferior surface of small wing of sphenoid, superior and anterior to optic foramen	Forms broad aponeurosis that splits into 3 lamellae: superficial blends with upper part of orbital septum and is prolonged forward above Tarsalis superior to deep surface of skin of superior eyelid; middle into upper margin of Tarsalis superior; deepest into superior fornix of conjunctiva
Superior rectus N: Oculomotor (Cr. 3)	Superior part of fibrous ring surrounding optic foramen on superior, medial, and inferior margins	Into sclera about 6 mm behind cornea, on superior aspect of eyeball
Inferior rectus N: Oculomotor (Cr. 3)	Inferior part of fibrous ring surrounding optic foramen on superior, medial, and inferior margins	Into sclera about 6 mm behind cornea
Medial rectus N: Oculomotor (Cr. 3)	Medial part of fibrous ring surrounding optic foramen on upper, medial, and lower margins	Into sclera on medial aspect of eyeball, about 6 mm behind cornea
Lateral rectus N: Abducent (Cr. 6)	Two heads from lateral parts of bands surrounding optic foramen and adjoining part of orbital fissure	Into sclera on lateral aspect of eyeball, about 6 mm behind cornea
Superior oblique N: Trochlear (Cr. 4)	Above margin of optic foramen, from body of sphenoid	Passes forward, ending in tendon that plays in fibrocartilaginous pulley attached to trochlear fovea of frontal bone; tendon passes backward, lateralward, and downward to lateral aspect of eyeball, inserting into sclera behind the equator of the eyeball; thus muscle pulls in a forward, upward, and medial direction
Inferior oblique N: Oculomotor (Cr. 3)	Orbital surface of maxilla, lateral to lacrimal groove	Passes lateralward, backward, and upward to insert into lateral part of sclera somewhat posterior to insertion of Superior oblique

MUSCLES OF THE MOUTH

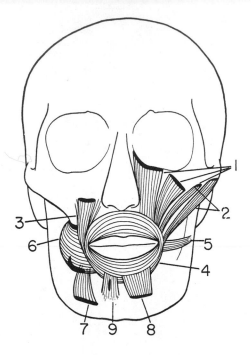

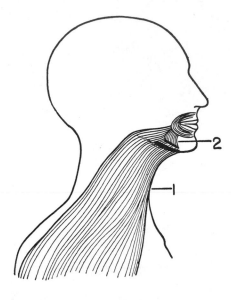

1. Levator labii superioris
2. Zygomaticus minor and major
3. Levator anguli oris
4. Obicularis oris
5. Risorius
6. Buccinator
7. Depressor anguli oris
8. Depressor labii inferioris
9. Mentalis

1. Platysma
2. Depressor anguli oris

PRIME MOVERS

Muscle	Origin	Insertion
Orbicularis oris N: Buccal branches of facial (Cr. 7)	a. Fibers derived from other facial muscles, principally Buccinator, Levator anguli oris, and Depressor anguli oris b. Proper fibers of lips, from under surface of skin c. Fibers attached to maxilla and septum of nose above and to mandible below	a. Intermingling of transverse and oblique fibers comprising muscle b. Mucous membrane lining mouth cavity c. Decussation of some fibers of Buccinator at corner of mouth; Levator anguli oris fibers pass below, and Depressor anguli oris fibers pass above mouth
Zygomaticus minor N: Buccal branches of facial (Cr. 7)	Malar surface of zygomatic bone posterior to zygomaticomaxillary suture	Upper lip between angular head and Levator anguli oris Upper lip at corner of mouth

MUSCLES OF THE MOUTH

ORBICULARIS ORIS

Patient approximates and compresses lips.

LEVATOR LABII SUPERIORIS AND ZYGOMATICUS MINOR

Patient protrudes and elevates upper lip.

LEVATOR ANGULI ORIS

Patient lifts upper border of lip on one side without raising lateral angle of mouth (sneering) (not illustrated).

ZYGOMATICUS MAJOR

Patient raises lateral angle of mouth upward and lateralward (smiling).

RISORIUS

Patient approximates lips and draws corners of mouth lateralward (grimacing).

BUCCINATOR

Patient approximates lips and compresses cheeks (blowing).

DEPRESSOR LABII INFERIORIS
AND MENTALIS

Patient raises the tip of chin and protrudes lower lip (pouting).

DEPRESSOR ANGULI ORIS AND PLATYSMA

Patient draws corners of mouth downward strongly.

MUSCLE	ORIGIN	INSERTION
Levator anguli oris N: Buccal branches of facial (Cr. 7)	Canine fossa, immediately below infraorbital foramen	Angle of mouth, intermingling with Zygomaticus, Depressor anguli oris, and Orbicularis
Zygomaticus major N: Buccal branches of facial (Cr. 7)	Zygomatic bone anterior to zygomaticotemporal suture	Angle of mouth, intermingling with Levator and Depressor anguli oris and Orbicularis oris
Risorius N: Mandibular and buccal branches of facial (Cr. 7)	Fascia over Masseter; muscle passes laterally superficial to Platysma	Skin at angle of mouth
Buccinator N: Buccal branches of facial (Cr. 7)	a. Outer surfaces of alveolar processes of maxilla above and mandible below, alongside the 3 molar teeth b. Pterygomandibular raphe	Fibers blend with deeper stratum of fibers in lips
Depressor anguli oris N: Mandible and buccal branches of facial (Cr. 7)	Oblique line of mandible	Angle of mouth
Depressor labii inferioris N: Buccal branches of facial (Cr. 7)	Oblique line of mandible, between symphysis and mental foramen	Skin of lower lip, blending with Orbicularis oris and opposite Depressor labii inferioris
Mentalis N: Mandibular and buccal branches of facial (Cr. 7)	Incisive fossa of mandible	Integument of chin
Platysma N: Cervical branch of facial (Cr. 7)	Fascia over superior Pectoralis major and Deltoid muscles	a. Anterior fibers interlace with opposite muscle inferior and posterior to symphysis menti b. Posterior fibers insert into mandible below oblique line or blend with muscles near angle of mouth

MUSCLES OF MASTICATION

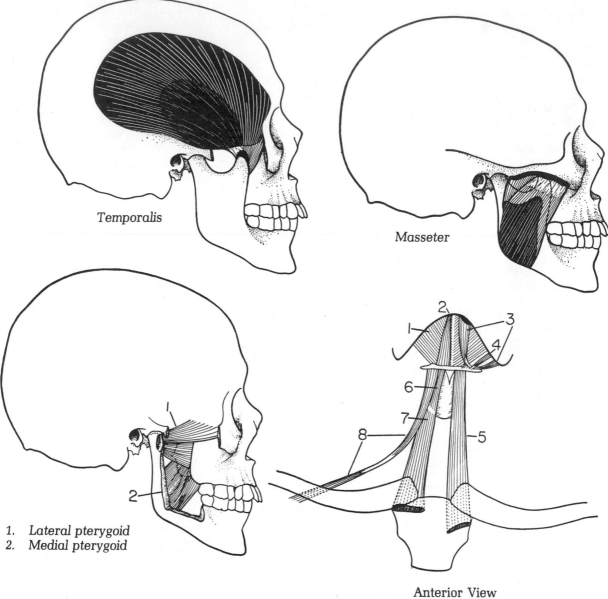

Temporalis

Masseter

1. Lateral pterygoid
2. Medial pterygoid

Anterior View

1.	Mylohyoid	5.	Sternohyoid
2.	Geniohyoid	6.	Thyrohyoid
3.	Digastric	7.	Sternothyroid
4.	Stylohyoid	8.	Omohyoid

PRIME MOVERS

MUSCLE	ORIGIN	INSERTION
Temporalis N: Deep temporal nerves from mandibular division of trigeminal (Cr. 5)	a. Temporal fossa b. Deep surface of temporal fascia	a. Medial surface, apex, and anterior border of coronoid process of mandible b. Anterior border of ramus of mandible nearly as far anterior as last molar tooth

(Continued on Page 164)

MUSCLES OF MASTICATION

TEMPORALIS, MASSETER AND MEDIAL PTERYGOID

Patient closes jaws tightly.

LATERAL AND MEDIAL PTERYGOIDS (LEFT)

Patient moves mandible laterally and forward to the right.

DIGASTRIC AND SUPRAHYOID MUSCLES

(Hyoid bone is fixed by infrahyoid muscles.) Patient depresses mandible.

MUSCLE	ORIGIN	INSERTION
Masseter N: Masseteric nerve from mandibular division of trigeminal (Cr. 5)	Superficial portion: a. Zygomatic process of maxilla b. Anterior two thirds of inferior border of zygomatic arch Deep portion: a. Posterior third of inferior border of zygomatic arch b. Whole medial surface of zygomatic arch	Angle and lower half of lateral surface of ramus of mandible a. Lateral surface of superior half of ramus of mandible b. Lateral surface of coronoid process
Lateral pterygoid N: Lateral pterygoid nerve from mandibular division of trigeminal (Cr. 5)	Superior head: a. Inferior part of lateral surface of great wing of sphenoid b. Infratemporal crest Inferior head: Lateral surface of lateral pterygoid plate	a. Depression in anterior part of neck of condyle of mandible b. Anterior margin of articular disk of temporomandibular articulation
Medial pterygoid N: Medial pterygoid nerve from mandibular division of trigeminal (Cr. 5)	a. Medial surface of lateral pterygoid plate b. Pyramidal process of palatine bone (second slip is lateral to lateral Pterygoid)	Inferior and posterior parts of medial surface of ramus and angle of mandibular foramen
Mylohyoid N: Inferior alveolar branch of trigeminal (Cr. 5)	Whole length of mylohyoid line of mandible, from symphysis in front to last molar tooth	Body of hyoid bone
Stylohyoid N: Facial (Cr. 7) For Muscles below, see innervation chart for anterolateral region of neck	Styloid process near its base	Body of hyoid bone at junction with greater cornu, just above Omohyoid muscle
Digastric N: Ant. belly, infer. alveolar branch of trigeminal (Cr. 5) Post. belly, facial (Cr. 7)	Anterior belly: Depression on inner side of inferior border of mandible Posterior belly: mastoid notch of temporal bone	(The two portions are united by an intermediate rounded tendon which perforates the Stylohyoid muscle)
Geniohyoid N: (C1) via hypoglossal (Cr. 12)	Inferior mental spine on inner surface of symphysis menti	Anterior surface of body of hyoid bone
Sternohyoid N: (C1, 2, 3) via ansa cervicalis	a. Posterior surface of medial end of clavicle b. Posterior and superior part of manubrium	Inferior border of body of hyoid bone
Thyrohyoid N: (C1) via hypoglossal (Cr. 12)	Oblique line on lamina of thyroid cartilage	Inferior border of greater cornu of hyoid bone
Sternothyroid N: (C1, 2, 3) via ansa cervicalis	Dorsal surface of manubrium below Sternohyoid, from edge of cartilage of first and sometimes second rib	Oblique line on lamina of thyroid cartilage
Omohyoid N: (C1, 2, 3) via ansa cervicalis	Inferior belly: cranial border of scapula, occasionally superior transverse ligament Superior belly: caudal border of body of hyoid bone	(The two portions are united by a central tendon held in position by a sheath of deep cervical fascia which is anchored to the clavicle and first rib)

II SCREENING THE AMBULATORY PATIENT FOR MUSCLE TESTING BY GAIT ANALYSIS

BACKGROUND OF GAIT ANALYSIS

The subject of human gait has been of interest since man first experienced limitations in locomotion. In 1680 Borelli had the distinction of being the pioneer in a recorded approach to the problems of human movement including locomotion. Little progress was made until the first half of the nineteenth century, however, when the observational period in gait analysis began with the brothers Wilhelm and Edward Weber. Steindler summarizes their contribution as the observation and measurement of "alternation of swing and support, the inclination of the trunk in either phase, the relationship between the duration and the length of the step, and the rhythm of alternation in walking and running, setting up a special pattern for each. They also initiated investigations into the muscle effort involved in propulsion and restraint."

Visual recording, using photography and kymography, followed and permitted recording phases of gait. It was then possible to calculate velocity, acceleration, and moving forces. The work of Braune and Fischer (1895) is the classic example of this approach.

More recently, study of individual muscles has permitted analysis of a pattern sequence of movement, first by palpation, later by electromyography. Morton and Schwartz added the concept and techniques of recording foot pressures.

The dynamic study of muscles encompassing the intensity and duration of muscle effort has provided still more tools for approaching the variations from normal gait so important to the clinician.

GAIT ANALYSIS

Phases of the Normal Pattern of Gait

There are two basic phases in the full cycle of a step: stance, the weight-bearing period; and swing, the nonweight-bearing portion of the step. In research on gait, a number of subdivisions of the two phases have been developed for study in detail; for example: the swing phase has been divided into initial swing, midswing and terminal swing. A number of such divisions are important in research but appear quite complex for a quick and accurate assessment by a busy clinician of an abnormality in gait. Instead of using such divisions, four major points in the cycle of a single step have been selected for easy identification and for recording any deviations from the normal gait. They are, in succession:

> Heel-strike
> Mid-stance
> Push-off
> Mid-swing

Elements of the Normal Pattern of Gait

Alignment
1. Head is erect.
2. Shoulders are level.
3. Trunk is vertical.

Cross Movements

1. Arms swing reciprocally and with equal amplitude at normal walking speed.
2. Steps are of the same length, and timing is synchronized.
3. Body undergoes vertical oscillations that are definite and have an even tempo.

Fine Movements

1. Pelvis undergoes slight
   a. transverse rotation: rotation is medial from the end of push-off to mid-stance and lateral from mid-stance through push-off.
   b. anterior-posterior tilt: forward inclination of the pelvis is maintained during the cycle with the exception of the stance phase at which time the pelvis tends to become level. The maximum amount of anterior tilt is evident before heel-strike and the minimum before mid-stance (excursion 3–5 degrees);
   c. list: maximum list is at mid-swing on the side of the swinging leg; and
   d. lateral displacement: maximum displacement is lateral at mid-stance on the side of the weight-bearing leg.
2. Hips rotate slightly medially during swing and heel-strike to near mid-stance, followed by a change to lateral rotation that continues through push-off.
3. Knees have two alternations of extension and flexion in a single cycle:
   a. extension of knee at heel-strike (not locked or tightly extended);
   b. slight flexion following heel-strike (continuing through mid-stance);
   c. extension following mid-stance; and
   d. flexion during push-off and swing.
4. Ankles
   a. rotate forward in an arc about a radius formed by the heel at heel-strike and around a center in the forefoot at push-off; and
   b. display maximum dorsiflexion at the end of stance phase and maximum plantar-flexion at the end of push-off.

Causes of Deviations in Gait

There are many factors that cause deviations in the normal pattern of walking. The most common are:

1. Pain or discomfort during weight-bearing or movement;
2. Muscle weakness;
3. Limitation of joint motion (often with muscle shortening);
4. Incoordination of movement; and
5. Changes in bone or soft tissue (including amputations).

Pain or *discomfort* can bring about distortions in the gait pattern varying from minor changes in alignment or movement to extreme deviations. Before observations are recorded, these factors should be explored by questioning the patient and noting protective responses. Clues such as shortening the stance (weight-bearing phase) or facial expression are of particular importance if the patient is unable to respond to questions.

Muscle weakness may be moderate and generalized, resulting in a broadened base of support, short steps, diminished arm swing, and difficulty in balance. In other instances, there may be extensive weakness in certain muscle groups but sufficient strength in others to allow ambulation. The extreme deviations are usually seen in the latter group.

Limitation of joint motion is most commonly found as a result of (1) a pathological condition—such as arthritis; (2) surgical procedures—for example, the insertion of metal implants; and (3) disuse of a part or of the whole body from a variety of causes. Deviations in gait due to joint limitation can be identified in the analysis and verified by joint measurements.

Incoordination resulting from neuropathological conditions (e.g. spastic cerebral palsy, hemiparesis secondary to cerebral vascular accident, Parkinson's syndrome), often leads to distinctive patterns of walking that can readily be identified and a description recorded. Hypertonic states are characterized by the lack of ability to activate muscle groups selectively and to combine them into the various patterns required for normal walking. The patient tends to respond with a total flexion or a total extension pattern when moving the limbs.

Deformities of bone and soft tissue lead to a variety of deviations in gait. Examples of such deformities are bone shortening following a fracture, congenital malformations, or dense scar tissue from a severe burn.

OBJECTIVES OF GAIT ANALYSIS

The objectives of gait analysis are to identify deviations and obtain information that may assist in determining the cause of the deviations and provide a basis for the use of therapeutic procedures or supportive devices to improve the walking pattern. Muscle tests can be used to determine the level of muscular weakness. However, by providing a gait analysis to screen ambulatory patients, it is possible to shorten and simplify the test procedures. Apparent areas of weakness found in the gait analysis may be validated by the results of the muscle tests and the accuracy of the tests supported by the findings in the analysis.

Procedure

A special sheet may be devised for recording the results of the gait analysis for patients with extensive involvement. Headings can be used such as Gait Deviation, Cause of Deviation, and Corrective Procedures. A patient's disability, however, is frequently limited to one area, and the information can be recorded without the use of a separate sheet.

The patient, if possible, should walk at a speed that is considered by the examiner to be normal for his age. A very slow gait tends to mask deviations as the steps are shortened and ranges of motion minimized. Such a gait may indicate generalized weakness with instability. If incoordination appears to be a problem, the ability to produce joint motion at hip, knee, and ankle in various combinations may be tested in order to determine the particular combination, or combinations, lacking for a normal gait.

After deviations in gait resulting from pain, generalized weakness, incoordination and fixed deformities have been screened out, the alignment of the body and gross movements should be reviewed. Any variation from the normal pattern indicates the need for a consideration of the fine movements with attention to each segment in sequence. If deviations are present, they are noted, and follow-up muscle tests or other evaluative procedures are instituted as indicated.

The next section is devoted to an analysis of (1) the four phases of normal gait with lateral and anterior or posterior views to illustrate each; and (2) the common deviations in gait with muscles that may be tested for each deviation. The range of motion to be checked is included where limitation is commonly the cause of the deviation.

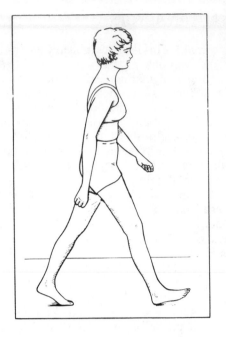

NORMAL PATTERN FROM *LATERAL* VIEW

(Right Leg)

1. Head and trunk are vertical. (Right arm is posterior to midline of body with elbow extended; left arm is anterior with elbow partially flexed.)
2. Pelvis has a slight anterior tilt.
3. Knee is extended but not locked.
4. Ankle is in slight plantar flexion.

Common Deviations from Normal

*1. Head and trunk are shifted forward.
2. Pelvis has a posterior tilt.

3. Knee is in locked extension or hyperextension.
4. Foot is placed flat on floor. There may be a slapping of forefoot.

Test the Following Muscles

1. Knee extensors.
2. Back extensors and hip flexors (check range of motion in hip flexion).
3. Knee extensors and flexors.

4. Ankle dorsiflexors.

*Places center of gravity anterior to knee joint to prevent knee flexion.

GAIT PHASE I: HEEL-STRIKE

NORMAL PATTERN FROM *ANTERIOR* VIEW

(Right Leg)

1. Head and trunk are vertical. (Arms swing at an equal distance from body.)
2. Pelvis lists slightly on right side.
3. Thigh and leg are in vertical alignment with pelvis.
4. Hip has slight medial rotation.
5. Plantar surface of forefoot is visible (plantar surface of heel would not be visible to a standing observer).

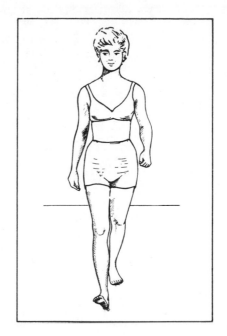

Common Deviations from Normal

*1. Trunk is displaced to right, and hip is in lateral rotation (step is shortened).
2. Hip is in abduction.
3. Plantar surface of forefoot is not visible.

Test the Following Muscles

1. Hip medial rotators, knee extensors and evertors of foot.
2. Hip adductors.
3. Ankle dorsiflexors.

---

*If knee extensors are weak, leg may be placed in lateral rotation at the hip to prevent flexion. If foot evertors are weak, lateral rotation prevents foot from rolling outward.

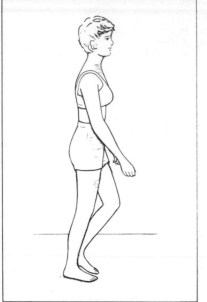

GAIT PHASE II: MID-STANCE

NORMAL PATTERN FROM *LATERAL* VIEW

(Right Leg)

1. Head and trunk are vertical. (Arms are near midline of body, elbows partially flexed.)
2. Pelvis has slight lateral rotation.
3. Knee is in slight flexion.
4. Ankle is in slight dorsiflexion.

Common Deviations from Normal	*Test the Following Muscles*
*1. Head and trunk are shifted forward at hip joint with exaggerated anterior tilting of pelvis.	1. Knee extensors.
†2. Head and trunk are shifted backward at hip joint with posterior tilting of pelvis.	2. Hip extensors.
3. Pelvis has exaggerated anterior tilting.	3. Abdominals and hip extensors (check range of motion in hip extension).
4. Knee is in locked extension or hyperextension.	4. Knee flexors and extensors and ankle dorsiflexors (check range of motion in dorsiflexion of ankle).
5. Knee has exaggerated flexion.	5. Ankle plantar flexors.
6. Ankle has exaggerated dorsiflexion (calcaneous position).	6. Ankle plantar flexors.

---

*Places center of gravity anterior to knee joint to prevent knee flexion.
†Places center of gravity posterior to hip joint to prevent forward shift of trunk.

GAIT PHASE II: MID-STANCE

NORMAL PATTERN FROM *ANTERIOR* VIEW

(Right Leg)

1. Head and trunk are vertical. (Arms are at an equal distance from body.)
2. Pelvis lists slightly on left side.
3. Hip has slight lateral rotation.

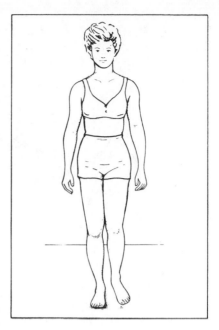

Common Deviations from Normal

*1. Head and trunk tip to right and pelvis tilts *upward* on left side. Right arm is away from body (Gluteus Medius gait).
2. Pelvis has an exaggerated list on left side (Trendelenburg gait).
3. Hip is in exaggerated outward rotation.

4. Foot is in varus position.
5. Foot is in valgus position.

Test the Following Muscles

1. *Right* hip abductors.

2. *Right* hip abductors.

3. Hip adductors, medial rotators, knee extensors, and foot evertors.
4. Foot evertors.
5. Foot invertors.

---

*If bilateral, patient tips to each side alternately with "waddle" gait.

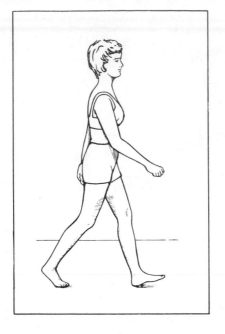

NORMAL PATTERN FROM *LATERAL* VIEW

(Right Leg)

1. Right arm is anterior to midline of body with elbow partially flexed, left arm is posterior with elbow extended.
2. Pelvis has an anterior tilt.
3. Knee is partially flexed.
4. Ankle is plantar flexed.
5. Toes are in extension at metatarsophalangeal joint.

Common Deviations from Normal

*1. Arms are at an uneven distance from midline body with both elbows flexed.
2. Pelvis has an exaggerated anterior tilt.

3. Hip may be laterally rotated and the knee forcedly extended as a skater does to propel the body forward.
4. Plantar flexion is limited, and ankle may be in dorsiflexion.
5. Metatarsophalangeal joints are straight.

Test the Following Muscles

1. Ankle plantar flexors and hip and knee extensors.
2. Abdominals and hip extensors (check range of motion in hip extension).
3. Ankle plantar flexors.

4. Same as Number 3.

5. Same as Number 3 (check range of motion in extension).

---

*An exaggerated forceful arm swing is used to assist in push-off, and step may be shortened.

GAIT PHASE III: PUSH-OFF

NORMAL PATTERN FROM *POSTERIOR* VIEW

(Right Leg)

1. Arms are at an equal distance from body, right elbow partially flexed, and left extended.
2. Hip is slightly laterally rotated.
3. Plantar surface of heel and midfoot are visible, and forefoot is in contact with floor.

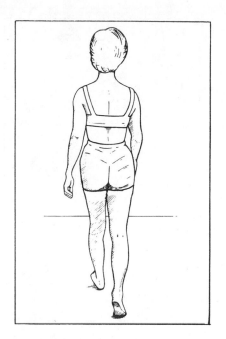

Common Deviations from Normal

*1. Arms are at an unequal distance from body with both elbows flexed.
†2. Hip is in exaggerated lateral rotation.
 3. Plantar surface of heel is not visible. Forefoot is not in contact with floor as heel is lifted.

Test the Following Muscles

1. Ankle plantar flexors and hip and knee extensors.
2. Same as Number 1.
3. Same as Number 1.

---

*See previous footnote.
†Knee may be forcefully extended to assist in push-off.

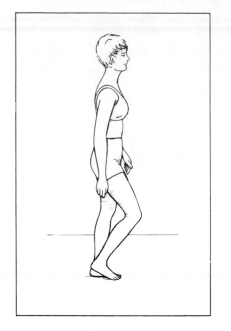

NORMAL PATTERN FROM *LATERAL* VIEW

(Right Leg)

1. Arms are near midline of body.
2. Pelvis has a very slight anterior tilt.
3. Hip and knee are flexed.
4. Ankle is in slight dorsiflexion.

Common Deviations from Normal

1. Pelvis has posterior tilting.

2. Hip and knee flexion are exaggerated and forefoot is dropped (Steppage gait).
3. Contralateral ankle has exaggerated plantar flexion.
4. Toes drag on floor.

Test the Following Muscles

1. Back extensors and hip flexors (check range of motion in hip flexion).
2. Ankle dorsiflexors.

3. Hip flexors, knee flexors, and ankle dorsiflexors.
4. Same as Number 3.

GAIT PHASE IV: MID-SWING

NORMAL PATTERN FROM *ANTERIOR* VIEW

(Right Leg)

1. Head and trunk are vertical.
2. Arms are at an equal distance from body.
3. Pelvis has a slight list on right side.
4. Hip is in slight medial rotation.
5. Ankle has slight eversion.

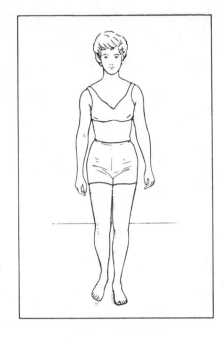

Common Deviations from Normal

1. Trunk is displaced to left. Pelvis is lifted on right side (hip hiking).
*2. Hip is in abduction.

3. Hip is laterally rotated.
4. Forefoot is dropped, eversion is not visible.

Test the Following Muscles

1. Hip and knee flexors and ankle dorsiflexors.
2. Same as Number 1 (check range of motion in hip adduction and flexion and knee flexion).
3. Hip medial rotators and foot evertors.
4. Ankle dorsiflexors and foot evertors.

---

*Leg may be circumducted (an arc of abduction) through swing phase.

REFERENCES

American Academy of Orthopaedic Surgeons. Joint Motion: Method of Measuring and Recording. Chicago, 1965.

Bailey, J. C. Manual Muscle Testing in Industry. Phys. Ther. Rev., 41:165–169, 1961.

Basmajian, J. V. Muscles Alive: Their Functions Revealed by Electromyography 5th ed. Baltimore, Williams & Wilkins Co., 1985.

Bennett, R. L. Muscle Testing: A Discussion of the Importance of Accurate Muscle Testing. Physiotherapy Rev., 27:242–243, 1947.

Borden, R., and Colachis, S. C. Quantitative Measurement of the Good and Normal Ranges in Muscle Testing. Phys. Ther., 48:839–843, 1968.

Borelli, J. A. De Motu Animalium. Rome, 1680.

Braune, C. W., and Fisher, O. Der Gang des Menschen. I. Teil. Versuche unbelasten und belasten Menschen. Abhandl. d. Math.-Phys. Cl. d. k. Sachs. Gesellsch. Wissensch., 21:153–322, 1895.

Brunnstrom, S. Muscle Group Testing. Physiotherapy Rev., 21:3–21, 1941.

Brunnstrom, S., and Dennen, M. Round Table on Muscle Testing. Annual Conference of American Physiotherapy Association, Federation of Crippled and Disabled, Inc., New York, 1931 (mimeographed).

Chusid, J. G. Correlative Neuroanatomy and Functional Neurology. 17th ed. Los Altos, Calif., Lange Medical Publications, 1979.

Daniels, L. Selected Methods of Grading Manual Muscle Tests with Suggestions for a Teaching Program. Thesis, Division of Physical Therapy, Stanford University, 1945.

Daniels, L. Measuring the Ranges of Joint Motion. Mimeographed Syllabus, Division of Physical Therapy, Stanford University, Revised 1972.

Downer, A. H. Strength of the Elbow Flexor Muscles. Phys. Ther. Rev., 33:68–70, 1953.

Ducroquet R., Ducroquet, J., and Ducroquet, P. Walking and Limping: A Study of Normal and Pathological Walking. Philadelphia, J. B. Lippincott Company, 1965. English Translation, 1968.

Eberhart, H. D., Inman, V. T., and Bresler, B. The Principal Elements in Human Locomotion. In: Klopsteg, P. E. and Wilson, P. D. Human Limbs and Their Substitutes, Chapter 15. New York, McGraw-Hill Book Company, 1954.

Eitner, D., Meissner, L., and Ork, H. Physical Therapy for Sports. Edited by W. Kuprian. Philadelphia, W. B. Saunders Company, 1982.

Ekstrand, J., Wiktorsson, M., Oberg, B., and Gillquist, J. Lower Extremity Goniometric Measurements; Study to Determine their Reliability. Arch. Phys. Med. Rehabil., 63:171–195, 1982.

Esch, D. and Lepley, M. Evaluation of Joint Motion: Methods of Measurement and Recording. Minneapolis, University of Minnesota Press, 1974.

Fisher, F. J., and Houtz, S. J. Evaluation of the Function of the Gluteus Maximus Muscle. Amer. J. Phys. Med., 47:182–191, 1968.

Gardner, E., Gray, D. J., and O'Rahilly, R. Anatomy: A Regional Study of Human Structure, 4th ed. Philadelphia, W. B. Saunders Company, 1975.

Gonnella, C. The Manual Muscle Test in the Patient's Evaluation and Program for Treatment. Phys. Ther. Rev., 34:16–18, 1954.

Granger, C. V. The Clinical Discernment of Muscle Weakness. Arch. Phys. Med., 44:430–438, 1963.

Gray, H. Anatomy of the Human Body. 30th ed. Edited by C. D. Clemente. Philadelphia, Lea & Febiger, 1984.

Gray's Anatomy. 36th British ed. Edited by R. Warwick and P. Williams. Philadelphia, W. B. Saunders Company, 1980.

Green, D. L., and Morris, J. M. Role of the Adductor Longus and Magnus in Postural Movements and in Ambulation. Amer. J. Phys. Med., 19:223–239, 1970.

Guide for Muscle Testing of the Upper Extremity. Downey, Calif., Occupational Therapy Department, Rancho Los Amigos Hospital, 1978.

Hallum, A. Goniometry. Mimeographed Syllabus, Division of Physical Therapy, Stanford University, Revised 1982.

Hines, T. F. Manual Muscle Examination. In: Therapeutic Exercise. Edited by S. Licht. New Haven, E. Licht Publisher, 1961.

Hoppenfeld, S. Physical Examination of the Spine and Extremities. New York, Appleton-Century-Crofts, 1976.

Iddings, D. M., Smith L. K., and Spencer, W. A. Muscle Testing: Part 2. Reliability in Clinical Use. Phys. Ther. Rev., 41:249–256, 1961

Inman, V. T. Functional Aspects of the Abductor Muscles of the Hip. J. Bone Joint Surg., 29:607–619, 1947.

Inman, V. T., Ralston, H. J., and Todd, F. Human Walking. Baltimore, Williams & Wilkins, 1981.

Janda, V. Muscle Function Testing. Boston, Butterworths, 1983.

Jarvis, D. K. Relative Strength of Hip Rotator Muscle Groups. Phys. Ther. Rev., 32:500–503, 1952.

Kapanji, I. A. The Physiology of the Joints. New York, Churchill Livingstone. Vol. 1, Upper Limb, 2nd English Edition, 1982; Vol. 2, Lower Limb, 1st English Edition, 1970.

Kendall, F. P., and McCreary, E. K. Muscles: Testing and Function. 3rd ed. Baltimore, Williams & Wilkins, 1983.

Kendall, H. O., and Kendall, F. P. Care During the Recovery Period in Paralytic Poliomyelitis. U.S. Public Health Bull. No. 242, revised, 1939.

LeVeau, B. Williams and Lissner Biomechanics of Human Motion. Philadelphia, W. B. Saunders Company, 1977.

Lilienfeld, A. M., Jacobs, M., and Willis, M. A Study of the Reproducibility of Muscle Testing and Certain Other Aspects of Muscle Scoring. Phys. Ther. Rev. 34:279–289, 1954.

Lovett, R. W., and Martin, E. G.: Certain Aspects of Infantile Paralysis and a Description of a Method of Muscle Testing. J.A.M.A., 66:729–733, 1916.

Lowman, C. L. A Method of Recording Muscle Tests. Amer. J. Surg., New Series 3:588–591, 1927.

Lowman, C. L. Muscle Strength Testing. Physiotherapy Rev., 20:69–71, 1940.

MacConaill, M. A., and Basmajian, J. V. Muscles and Movements, a Basis for Human Kinesiology, 2nd ed. Baltimore, Williams & Wilkins Company, 1977.

McMahon, T. A. Muscles, Reflexes and Locomotion. Princeton University Press, 1984.

Moore, M. L. Clinical Assessment of Joint Motion. In: Licht, S. Therapeutic Exercise. Baltimore, Waverly Press, Inc., 1965.

Murray, M. P. Walking Patterns of Normal Women. Arch. Phys. Med. Rehab., 51:637–650, 1970.

Murray, M. P., Drought, A. B., and Kory, R. C. Walking Patterns of Normal Men. J. Bone Joint Surg., 46-A:335–360, 1964.

Norkin, C. and White, J. Measurement of Joint Motion: A Guide to Goniometry. Philadelphia, F.A. Davis Co., 1985.

Partridge, M. J., and Walters, C. E. Participation of the Abdominal Muscles in Various Movements of the Trunk in Man: An Electromyographic Study. Phys. Ther. Rev. 39:791–800, 1959.

Perry, J. A. Clinical Interpretation of the Mechanics of Walking. Phys. Ther., 47:778–801, 1967.

Perry, J. Normal Upper Extremity Kinesiology. Phys. Ther., 58(3):265–278, 1978.

Plastridge, A. L. Gaits. Physiotherapy Rev. 21:24–29, 1941.

Plastridge, A. L. Round Table on Infantile Paralysis. Annual Conference of American Physiotherapy Association. Stanford University. Stanford University Press, 1942.

Pocock, G. S. Electromyographic Study of the Quadriceps During Resistive Exercise. J. Amer. Phys. Ther. Assoc., 43:427–434, 1963.

Reeder, T. Electromyographic Study of the Latissimus Dorsi Muscle. J. Amer Phys. Ther. Assoc., 43:165–172, 1963.

Rodenberger, M. L.: Method of Recording Progress in Gait Training. Phys. Ther. Rev., 30:92–94, 1950.

Saunders, J. B. DeC. M., Inman, V. T., and Eberhart, H. D. The Major Determinants in Normal and Pathological Gait. J. Bone Joint Surg., 35-A:543–558, 1953.

Schwartz, R. P., Heath, A. I., Morgan, D. W., and Towne, R. C. A Quantitative Analysis of Recorded Variables in the Walking Patterns of "Normal" Adults. J. Bone Joint Surg., 46:324–334, 1964.

Smith, L. K., Iddings, D. M., Spencer, W. A., and Harrington, P. R. Muscle Testing, Part I. Description of a Numerical Index for Clinical Research. Phys. Ther. Rev., 41:99–105, 1961.

Steindler, A. Historical Review of the Studies and Investigations Made in Relation to Human Gait. J. Bone Joint Surg., 35-A:540–543, 1953.

Sutherland, D. Gait Disorders in Childhood and Adolescence. Baltimore, Williams & Wilkins Company, 1984.

Sutherland, D., Cooper, L. and Daniel, D. Role of the Ankle Plantar Flexors in Normal Walking. J. Bone Joint Surg., 62A:354–363, 1980.

Weber, W., and Weber, E. Mechanik der menschlichen Gehwerkzeuge, Gottingen, Dietrich, 1836.

Williams, M. Manual Muscle Testing: Development and Current Use. Phys. Ther. Rev., 36:797–805, 1956.

Williams, M., and Lissner, H. R. Biomechanical Analysis of Knee Function. J. Amer. Phys. Ther. Assoc., 43:93–99, 1963.

Williams, M., and Stutzman, L. Strength Variation Through the Range of Joint Motion. Phys. Ther. Rev. 39:145–152, 1959.

Wintz, M. M. Variations in Current Muscle Testing. Phys. Ther. Rev. 39:466–475, 1959.

Worthingham, C. A. Upper and Lower Extremity Muscle and Innervation Charts. Stanford, Stanford University Press, 1944.

Wright, W. G. Muscle Training in the Treatment of Infantile Paralysis. Boston, M. & S.. J., 167:567–574, 1912.

INDEX

Page numbers in *italics* indicate illustrations. Page numbers followed by (t) indicate tables.

Muscle(s) *(Continued)*

  innervation of. See *Muscle innervation.*
  internal abdominal oblique, as accessory, 22(t), 34(t)
    in trunk rotation, 26-28, *26,* 26(t)
  lateral cervical, neck, innervation, 14(t)
  lateral pterygoid, 162-164, *162, 163,* 164(t)
  lateral vertebral, neck, in muscle innervation, 14(t)
  lateral rectus, 154-156, *154, 155,* 156(t)
  latissimus dorsi, as accessory, 26(t), 34(t)
    in shoulder extension, 106-107, *106,* 106(t)
    in shoulder rotation, medial, 116-117, 116(t)
  levator anguli oris, 158-161, *158, 159,* 161(t)
  levator labii superioris, 158-161, *158, 159*
  levator palpebrae superioris, 154-156, *154, 155,*
    156(t)
  levator scapulae, as accessory, 20(t)
    in scapular elevation, 92-93, *92,* 92(t)
  longissimus capitis, in neck extension, 20(t)
  longissimus cervicis, in neck extension, 20(t)
  longissimus thoracis, in trunk extension, *30,* 30(t)
  longus capitis, as accessory, 16(t)
  longus colli, as accessory, 16(t)
  lumbricals, in flexion, metacarpophalangeal joints,
    130-131, *130,* 130(t)
    in toe flexion, metatarsophalangeal joints, 82-83,
    82, 82(t)
  masseter, 162-164, *162, 163,* 164(t)
  medial pterygoid, 162-164, *162, 163,* 164(t)
  medial rectus, 154-156, *154, 155,* 156(t)
  mentalis, 158-161, *158, 160,* 161(t)
  multifidus, as accessory, 20(t), 26(t), 32(t)
  mylohyoid, 162-164, *162,* 164(t)
  nasalis, 152-153, *152,* 152(t)
  obliquus capitis inferior, as accessory, 20(t)
  obliquus capitis superior, as accessory, 20(t)
  obturator externus, in hip lateral rotation, 58-60,
    58, 58(t)
  obturator internus, in hip lateral rotation, 58-60,
    58, 58(t)
  occipitofrontalis, 152-153, *152,* 152(t), *153*
  of eye, 154-156, *154, 155,* 155(t), 156(t)
    innervation of, 150(t)
  of facial expression, innervation of, 150(t)
  of forehead, 152-153, *152,* 152(t), *153*
  of mastication, 162-164, *162,* 162(t), *163,* 164(t)
    innervation of, 150(t)
  of mouth, 158-161, *158,* 158(t), *159, 160,* 161(t)
  of nose, 152-153, *152,* 152(t), *153*
  omohyoid, 162-164, *162,* 164(t)
  opponens digiti minimi, in fifth finger-thumb
    opposition, 148-149, *148,* 148(t)
  opponens pollicis, in fifth finger-thumb opposition,
    148-149, *148,* 148(t)
  orbicularis oculi, 154-156, *154, 155,* 155(t)
  orbicularis oris, 158-161, *158,* 158(t), *159*
  palmar interossei, in finger adduction, 138-139,
    138, 138(t)
    in flexion, metacarpophalangeal joints, 130-131,
    130, 130(t)
  palmaris longus, as accessory, 126(t), 144(t)
  pectineus, as accessory, 38(t)
    in hip adduction, 54-56, *54,* 56(t)
  pectoralis major, as accessory, 102(t)
    in shoulder horizontal adduction, 112-113, *112,*
    112(t)
    in shoulder rotation, medial, 116-117, 116(t)
  peroneus brevis, as accessory, 72(t)
    in foot eversion, 80-81, *80,* 80(t)
  peroneus longus, as accessory, 72(t)
    in foot eversion, 80-81, *80,* 80(t)
  peroneus tertius, as accessory, 80(t)
  piriformis, in hip lateral rotation, 58-60, *58,* 58(t)
  plantar interossei, as accessory, 82(t)

Muscle(s) *(Continued)*

  plantaris, as accessory, 72(t)
  platysma, 158-161, *158, 160,* 161(t)
  popliteus, as accessory, 66(t)
  procerus, 152-153, *152,* 152(t), *153*
  pronator quadratus, in forearm pronation, 124-125,
    124, 125(t)
  pronator teres, in forearm pronation, 124-125, *124,*
    125(t)
  psoas major, in hip flexion, 38-40, *38,* 38(t)
  quadratus femoris, in hip lateral rotation, 58-60,
    58, 58(t)
  quadratus lumborum, in elevation of pelvis, 34-35,
    34, 34(t)
    in trunk extension, 32(t)
  quadriceps femoris, in knee extension, 68-70, *68,*
    68(t)
  rectus abdominis, as accessory, 26(t)
    in trunk flexion, 22-24, *22,* 22(t)
  rectus capitis anterior, as accessory, 16(t)
  rectus capitis posterior, as accessory, 20(t)
  rectus femoris, as accessory, 38(t)
    in knee extension, 68-70, *68,* 68(t)
  rhomboid major, as accessory, 92(t)
    in scapular adduction, 94-96, 94(t)
      with downward rotation, 100-101, *100,* 100(t)
  rhomboid minor, as accessory, 92(t)
    in scapular adduction, 94-96, 94(t)
      with downward rotation, 100-101, *100,* 100(t)
  risorius, 158-161, *158, 159,* 161(t)
  rotatores, as accessory, 26(t), 32(t)
  sacrospinalis, in neck extension, 20(t)
    in trunk extension, *30,* 30(t)
  sartorius, as accessory, 38(t), 60(t), 66(t)
    in combined joint motion, 42-43, *42,* 42(t)
  scalenus anterior, as accessory, 16(t)
  scalenus medius, as accessory, 16(t)
  scalenus posterior, as accessory, 16(t)
  semimembranosus, as accessory, 62(t)
    in hip extension, 44-46, *44,* 46(t)
    in knee flexion, 64-66, *64,* 66(t)
  semispinalis, as accessory, 26(t), 32(t)
  semispinalis capitis, in neck extension, 18-19, *18,*
    18(t)
  semispinalis cervicis, in neck extension, 20(t)
  semitendinosus, as accessory, 62(t)
    in hip extension, 44-46, *44,* 44(t)
    in knee flexion, 64-66, *64,* 64(t)
  serratus anterior, as accessory, 108(t)
    in abduction and upward rotation of scapula,
    90-91, *90,* 90(t)
  serratus anterior, as accessory, 102(t)
  soleus, in ankle plantar flexion, 72-74, *72,* 72(t)
  spinalis capitis, in neck extension, 20(t)
  spinalis cervicis, in neck extension, 20(t)
  spinalis thoracis, in trunk extension, *30,* 32(t)
  splenius capitis, in neck extension, 18-19, *18,* 18(t)
  splenius cervicis, in neck extension, 18-20, *18,* 20(t)
  sternocleidomastoid, in neck flexion, 16-17, *16,* 16(t)
  sternohyoid, 162-164, *162,* 164(t)
  sternothyroid, 162-164, *162,* 164(t)
  stylohyoid, 162-164, *162,* 164(t)
  suboccipital, innervation, 15(t)
  subscapularis, in shoulder rotation,
    medial, 116-117, *116,* 116(t)
  superficial, neck, innervation, 14(t)
  superior oblique, 154-156, *154, 155,* 156(t)
  superior rectus, 154-156, *154, 155,* 156(t)
  supinator, in forearm supination, 122, 122(t), *123*
  suprahyoids, neck, innervation, 14(t)
  supraspinatus, in shoulder abduction, 108-109, *108,*
    108(t)
  temporalis, 162-164, *162,* 162(t), *163*

184